Groping in the dark

Groping in the dark

The first decade of global modelling

DONELLA MEADOWS
JOHN RICHARDSON
GERHART BRUCKMANN

JOHN WILEY & SONS

Chichester · New York · Brisbane · Toronto · Singapore

Library of Congress Cataloging in Publication Data:

Meadows, Donella H.
Groping in the dark.

Includes index.
1. Economic history — Mathematical models.
I. Richardson, John. II. Bruckmann, Gerhart.
III. Title.
HC59.M413 330.9 81-14713

ISBN 0 471 10027 7 AACR2

British Library Cataloguing in Publication Data:

Groping in the dark.
1. Economic history– Mathematical models
I. Meadows, Donella
II. Richardson, John
III. Bruckmann, Gerhart
330.9 HC59

ISBN 0 471 10027 7

Typeset by Photo-Graphics, Yarcombe, Honiton, Devon, England.
Printed by J.W. Arrowsmith Limited, Bristol.

This Book Is Dedicated to

EVERYBODY
who is
NOT
GROPING IN THE DARK

A personal introduction

This is a message from the editors

Donella Meadows
John Richardson
Gerhart Bruckmann

Who want to tell you, in a personal way,

About how this book came to be written,
How we feel about what it says,
How it is organized,
And, most important,

Why you should take time from

Reading other books,
working
thinking
talking
sleeping
eating
loving
or whatever

To read it.

This book began as the official CONFERENCE PROCEEDINGS
of the Sixth IIASA Symposium on Global Modeling.
(For those of you who may not have heard of IIASA, it is the International Institute for Applied Systems Analysis in Laxenburg, Austria. This Institute concerns itself with large-scale problems, and therefore has been interested in global modelling since it was founded in 1972.)

It may strike you that *Groping in the Dark*
Does not exactly resemble other CONFERENCE PROCEEDINGS,
Which are

VERY VALUABLE BOOKS because they

Compile IMPORTANT IDEAS of IMPORTANT PEOPLE
Provide a HISTORICAL RECORD of what transpired

Offer DIVERSE PERSPECTIVES on a COMMON PROBLEM
Reflect, more or less, the STATE OF THE ART

But

Unfortunately,

CONFERENCE PROCEEDINGS are also often

Disorganized
Unfocused
Of uneven quality
Hastily compiled
Somewhat boring, and
Unread.

And they never have titles like *Groping in the Dark*.

Every editor and author hopes that his or her book will be
EXCITING, SIGNIFICANT, DIFFERENT, AND READ BY MILLIONS.
This book may not be

READ BY MILLIONS

But we are pretty sure that it is

DIFFERENT

And we think that it is

EXCITING AND SIGNIFICANT

Because it documents the struggle of some of the most interesting people we know to solve some of the most important problems we can think of.

(It's not often that you get to listen to very good minds from around the world trying to improve the course of human history
And being honest about how hard it is.)

We were EXCITED about this Conference before it began,
and even more so after it was over

So

We expanded the record of its Proceedings
to make it accessible to a wider audience
than just those who were there.

There are three kinds of people in this wider audience:

AUDIENCE # 1. Those who make computer models of complex social systems.
(The Conference participants and their colleagues)

AUDIENCE # 2. Those who are formally responsible for making complex social systems work.
(The decision makers and policy formulators we hear so much about)

AUDIENCE # 3. Those who live in and care about the complex social systems that the modellers model and the decision makers decide about.

In the case of global models, Audience # 3 consists of

everyone.

We thought a long time about how to make this book
Meet the needs of three different audiences and at the same time reflect
Our own feeling about the special importance of its contents.

Finally we remembered a book that was
Different from anything we had ever read before.
It is called

PLATFORM FOR CHANGE

and is by Stafford Beer, the British cybernetician and philosopher.
(If you haven't read it, you should.)

Remembering that book helped us design this one.

Like *Platform for Change*
This book is printed
On pages of different colors,
Each of which means something.

YELLOW PAGES (like this one) are for our personal thoughts, experiences, and reflections. Every author or editor has these, of course, but they are usually filtered out before publication.

Too bad, we think.

BLUE PAGES are for important conclusions and highlights from the Conference.

When we put something on a blue page it means that

We think it is important,
and a lot of other people think so too.

WHITE PAGES are for the actual proceedings of the Conference.
That is,

what everyone else (including ourselves when we were playing the role of everyone else) said.

GOLD PAGES introduce the field of global modelling and its history. They are intended to help many readers, especially Audiences #2 and #3, understand better what is written on the yellow, blue, and white pages.

This format may make you uncomfortable.

If so, you could ignore the yellow, blue, and gold pages,

In which case the book will be much like
All CONFERENCE PROCEEDINGS you have read
in the past.

Or

You could simply go back to

Reading other books,
working
thinking
talking
etc.

We hope you won't, because

We believe

that the future is going to be quite different from the past, that we all need to make decisions quite different from those we have made in the past,

that we need to start looking at the world in a new way, and

that

THIS BOOK IS ABOUT A NEW AND NEEDED WAY OF LOOKING AT THE WORLD

On the next page
we shall begin
by telling you the story that made us decide
to entitle this book

GROPING IN THE DARK

A fable from the past

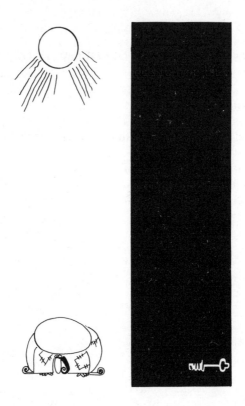

There is more Light here

Someone saw Nasrudin searching for something on the ground.

'What have you lost, Mulla?' he asked. 'My key,' said the Mulla. So they both went down on their knees and looked for it.

After a time the other man asked: 'Where exactly did you drop it?'

'In my own house.'

'Then why are you looking here?'

'There is more light here than inside my own house.'

A fable from the very recent past

Once upon a time, in the eighth decade of the twentieth century A.D., there was a global modeller by the name of G. He was unhappy. Often at night he lay awake, his eyes open. He was considered a good teacher, capable of communicating complicated ideas to an uninformed or uninterested audience. But, alas, when it came to the concept of global modelling, a subject dear to his heart, he seemed to fail to communicate, over and over again.

He was particularly frustrated when he discussed his craft with a colleague by the name of A, who was also a distinguished scholar and proficient modeller. Both G and A observed a strange phenomenon in their discussions. The words each one used obviously belonged to the English language, their syntax was in order, but somehow the sentences seemed to fly past each other without making any contact or conveying any meaning.

Finally, after an especially sleepless night, G resolved to try again. He invited A for a long walk in the quiet and beautiful park that adjoined their research institute. Once again G explained to A the interconnected complex of the world's problems, the need to operate with a comprehensive picture rather than a partial one, the difficulty of understanding the situation intuitively, and the urgency of taking some action with imperfect models rather than waiting for better ones that might come too late. A listened, carefully and silently. Then A expressed his concerns about the lack of rigour, sparse data, unwarranted assumptions, inadequate validation, and general naïveté that, in his view, characterized global modelling work. To drive his points home, A cited several contrasting examples from his own more rigorous, well-established, and focused field of enquiry.

When A had finished, the two men walked quietly for a time. They had almost reached the back door of the Institute again.

Finally G broke the silence. He said, 'For months we have tried in vain to understand where and how we disagree. Sometimes an ancient fable is more powerful than modern scientific language.' He then told A the story about the man searching for his lost key. He concluded:

'The key both you and I are trying to find is the solution to the critical problems mankind will face in the coming decades. Each of us is searching with sincerity and devotion. What is profoundly different, however, is our basic strategy. You stand in the light, trying to move the lightpost closer to the place where the key might be. I, on the other hand, am

GROPING IN THE DARK.'

Contents

Highlights

In the next few pages we summarize important ideas
From the rest of the book
For the three audiences to whom the book is directed.

For Audience # 1 (modellers) there are conclusions from the Conference about the practice of global modelling. A longer version is in Chapter 6.

These conclusions will also be useful to anyone who wants to ask modellers probing questions about the credibility of their models. They provide good clues about where the modellers are most and least certain.

For Audience # 2 (users) there is a summary of what modelling is all about, what one might be able to learn from it, and how to use it. There is more on this in Chapter 2.

This part may also be a helpful reminder to modellers about the essence of what they do, and a guide to everyone about the basis for all decisions, global or personal.

For Audience # 3 (everyone) there is a list of the conclusions about the world that can be derived from the global models so far. This list is elaborated at the end of Chapter 2.

Since this audience is the largest and this list is the most generally interesting, we shall begin the highlights there and work backwards.

Lessons about the World

Seven global models have been completed so far.

They have been made by different groups of people in different countries asking different questions and using different methods. Some of the models have been made deliberately to refute others.

The fact that all the global modellers can agree on anything is worthy of note.

In spite of their differences, it is possible to draw from the seven global models some common, general messages about the state of the world and its possible futures.

Here are the most important ones:

1. There is no known physical or technical reason why basic needs cannot be supplied for all the world's people into the foreseeable future. These needs are not being met now because of social and political structures, values, norms, and world views, not because of absolute physical scarcities.
2. Population and physical (material) capital cannot grow forever on a finite planet.
3. There is no reliable, complete information about the degree to which the earth's physical environment can absorb and meet the needs of further growth in population and capital. There is a great deal of partial information, which optimists read optimistically and pessimists read pessimistically.
4. Continuing 'business-as-usual' policies through the next few decades will not lead to a desirable future— or even to meeting basic human needs; it will result in an increasing gap between the rich and the poor, problems with resource availability and environmental destruction, and worsening economic conditions for most people.
5. Because of these difficulties, continuing current trends is not a likely future course. Over the next three decades the world socioeconomic system will be in a period of transition to some state that will be, not only quantitatively but also qualitatively, different from the present.
6. The exact nature of this future state, and whether it will be better or worse than the present, is not predetermined, but is a function of decisions and changes being made now.
7. Owing to the momentum inherent in the world's physical and social processes, policy changes made soon are likely to have more impact with less effort than the same set of changes made later. By the time a problem is obvious to everyone, it is often too far advanced to be solved.
8. Although technical changes are expected and needed, no set of purely technical changes tested in any of the models was sufficient in itself to

bring about a desirable future. Restructuring social, economic, and political systems was much more effective.

9. The interdependencies among peoples and nations over time and space are greater than commonly imagined. Actions taken at one time and on one part of the globe have far-reaching consequences that are impossible to predict intuitively, and probably also impossible to predict (totally, precisely, maybe at all) with computer models.
10. Because of these interdependencies, single, simple measures intended to reach narrowly defined goals are likely to be counterproductive. Decisions should be made within the broadest possible context, across space, time, and areas of knowledge.
11. Cooperative approaches to achieving individual or national goals often turn out to be more beneficial in the long run to all parties than competitive approaches.
12. Many plans, programmes, and agreements, particularly complex international ones, are based upon assumptions about the world that are either mutually inconsistent or inconsistent with physical reality. Much time and effort is spent designing and debating policies that are, in fact, simply impossible.

In short: CHANGE IN THE STATUS QUO IS CERTAIN. IMPROVEMENT IN THE STATE OF THE WORLD IS BY NO MEANS IMPOSSIBLE AND BY NO MEANS GUARANTEED. WE ARE A LONG WAY FROM KNOWING EVERYTHING WE NEED TO KNOW. AND YET WE KNOW ENOUGH ABOUT WHERE WE WANT TO GO AND HOW TO GET THERE TO GET STARTED. THE SITUATION IS NOT HOPELESS; IT IS CHALLENGING.

Lessons about Modelling

All human decisions are based on generalizations or abstractions about the world, or *models*. Most of these models exist only in people's heads. We call them *mental models*. Mental models are complex, shifting, and often unverbalized.

In order for a mental model to be communicated, it must be expressed in some medium of communication. *Words* are the most common medium. We call a model expressed in words a *verbal model*. Books like Aristotle's *Politics* or Marx's *Das Kapital* and sayings like 'to every thing there is a season and a time to every purpose under the heaven' are examples of verbal models. Different people reading the same verbal model can create very different mental models of what it means.

Mathematical models are also based on mental models, but are written down in the form of mathematical symbols. *Computer models* are mathematical models that have been written in computer language so that a computer can calculate the conclusions that follow from the assumptions. In general, a computer model is used when the situation or system being modelled is so complicated that it cannot be encompassed in words or simple equations. All of the global models discussed in this book are both mathematical models and computer models. We will usually refer to them as computer models.

There are many different types of computer models. Like mental models, they can be used for a variety of purposes:

- exact prescriptions of what to do;
- conditional forecasts about what might happen if . . .;
- general frameworks for understanding how systems work;
- organization of many scattered pieces of information;
- explicit presentation of a particular point of view.

EVERY DECISION IS BASED ON AN APPROXIMATION OF REALITY, NOT REALITY ITSELF. THE QUESTION IS NOT, 'SHOULD MODELS BE USED?' BUT, RATHER, 'WHICH IS THE BEST MODEL AVAILABLE FOR THE TASK AT HAND?'

Mental models are often more useful than computer models because they

- include rich stores of information about intangibles such as motivations and values;
- may be free from artificial boundaries imposed by academic disciplines or tradition (but they may be constrained by other, equally artificial, boundaries);
- can make creative associations and analogies;
- can strike a balance between incommensurable quantities and conflicting values;
- are readily available, quick, and cheap.

Computer models may sometimes be more useful than mental models because they

- are rigorous, precise, and consistent;
- are written out explicitly so that they can be understood and criticized by anyone;
- can contain many more variables and keep track of them all simultaneously;
- always draw error-free logical conclusions from their assumptions;
- can be changed and tested very quickly.

Computer models (like mental models) are not always as good as they could be, because of mistakes that modellers make. For instance, computer modellers sometimes

— make their models so complicated that no one, including themselves, can understand them;
— describe their models in such specialized jargon that no one can understand them;
— leave out things that are important but mathematically inconvenient;
— type things wrong, calculate inputs wrong, interpret outputs wrong, and make other typical human mistakes;
— fail to test their models thoroughly enough to find out whether they have made any mistakes.

All of these are human (and correctable) faults of modellers, not intrinsic faults of computer modelling.

MORE CAN BE LEARNED FROM APPROPRIATE AND SKEPTICAL USE OF BOTH MENTAL AND COMPUTER MODELS THAN FROM RELIANCE ON EITHER ONE ALONE.

Lessons about Global Modelling Practice

Gobal models are nothing more (or less) than computer models that address social problems of global scope.

Examples of such problems:

— resource depletion,
— poverty and hunger,
— inequities in international trade,
— environmental degradation,
— rapid population growth.

THE METHODOLOGICAL STRENGTHS AND WEAKNESSES OF GLOBAL MODELS ARE THE SAME AS THOSE OF OTHER POLICY-ORIENTED COMPUTER MODELS.

The global modellers at the Sixth Global Modelling Conference agreed about some points of modelling practice and agreed to disagree about others. The points of disagreement indicate where the major difficulties and uncertainties in the modelling process are. The points of agreement provide a set of guidelines for good modelling practice.

The areas of disagreement are these:

1. Should models be built to answer a single, well-defined problem or to represent many aspects of a system and serve many different purposes?
2. Should models be made in direct response to pressing issues of public policy or should the goal be general improvement in understanding?
3. Should models be normative or descriptive?
4. How far into the future can one really see with a model?
5. What is the best method to use for global modelling?
6. Should models be big or small?
7. Should the procedure for developing the model be top down (beginning with general, comprehensive structures and then adding detail) or bottom up (beginning with detailed pieces and assembling them into a total structure)?
8. What should be done when data about a crucial aspect of the system are not available?
9. How should actors, technology, prices, population, etc, be represented?
10. How should a model be tested?
11. What is the appropriate audience for global modelling? When and how should results be communicated to this audience?

The areas of agreement are these:

1. It is better to state your biases, insofar as you are able, than to pretend you don't have any.
2. Computer models of social systems should not be expected to produce precise predictions.
3. Inexact, qualitative understanding can be derived from computer models and can be useful.
4. Methods should be selected to fit problems (or systems); problems should not be distorted to fit methods.
5. The most important forces shaping the future are social and political, and these forces are the least well represented in the models so far.
6. In long-term global models, environmental and resource considerations have been too much ignored.
7. Models should be tested much more thoroughly than they typically are, for agreement with the real world, for sensitivity to uncertainties, and over the full range of possible policies.
8. A substantial fraction of modelling resources should go to documentation.
9. Part of the model documentation should be so technically complete that any other modelling group can run and explore the model and duplicate all of the published results.

10. Part of the documentation should be so clear and free from jargon that a nontechnical audience can understand all the model's assumptions and how these assumptions lead to the model's conclusions.
11. Modellers should identify their data sources clearly and share their data as much as possible.
12. Model users, if there are any clearly identifiable ones, should be involved in the modelling process as directly and frequently as possible.
13. An international clearinghouse for presenting, storing, comparing, criticizing, and publishing global models is necessary.

AS LONG AS THERE ARE GLOBAL PROBLEMS, THERE WILL BE A NEED FOR GLOBAL MODELS. MANY MODELS OF GLOBAL SCOPE, ADDRESSING DIFFERENT PROBLEMS FROM A VARIETY OF PERSPECTIVES, ARE NEEDED.

The Editors' biases

One of the major conclusions of this book is that
IT IS BETTER TO STATE YOUR BIASES THAN TO PRETEND YOU DON'T HAVE ANY.

So we decided to be sure that our biases,
at least those that may be important to shaping this book,
are clearly stated.

It is not an easy task to discover one's biases.

After contemplating our biases for a while,
the three of us, who know each other pretty well,
decided each should write down the biases of the OTHER TWO.

That was an amazing exercise—try it sometime.

We have dozens, maybe hundreds, of biases.
Some are observable or readily deducible from our habits, possessions, or backgrounds.
Others are harder to see, but no less important.
In the interest of brevity, we have edited our
biases, Here are some of them.

ALL THREE OF US

— are militantly interdisciplinary
— care about and worry about the long-term future of mankind
— value nature and the arts
— tend to be workaholics
— see the world through the paradigm that there are paradigms
— believe that the world is made of complex systems that we can't understand intuitively

— find the idea of global modelling exciting, but have reservations about the models made so far

DM

— is white, female, American, and academic
— was trained as a biophysicist, but now likes philosophy better
— dislikes big centralized systems, including modern capitalism and communism
— favours system dynamics as a modelling method
— lives in a commune, raises organic vegetables, jogs, and heats with wood
— studies Zen Buddhism, and therefore knows that there is such a thing as Truth, that it can be found within oneself and not in the world of illusion, and that it is not easy to find
— believes that all people are truly magnificent, that the world is a precious miracle, that the future could hold either unspeakable horrors or undreamed-of wonders, and that it is all up to us

JR

— is white, male, American, and academic
— was trained in history, political science, anthropology, and mathematics, but now teaches international affairs
— uses several different types of modelling methods
— abhors extremes in values, methods, and personal lifestyles
— lives on a farm and raises horses
— dislikes dogmatism and hypocrisy, especially dogmatism about being non-dogmatic and hypocrisy about being non-hypocritical
— accepts that there may be something called Truth, but is suspicious of anyone who claims to have found it
— believes that the future can be changed, and that he can play a role in changing it, but recognizes that he could be wrong on both counts

GB

— is white, male, academic, and very Viennese
— majored in seven different fields, ranging from engineering to economics, at five universities, but now makes a living by teaching statistics
— has never made a global model himself, but has presided at the public presentation of most of them
— loves rationality for its irrational beauty
— lives in a stylishly furnished apartment in Vienna

- values adaptation and change and at the same time appreciates form and tradition
- knows that if there is no Truth, this statement can also not be true
- sees life as a game that is tremendous fun to play, and believes that to help improve the future of mankind is part of the game

This list is only a shallow model of our biases.

The really big ones are the ones we don't see.

You will spot some additional ones as you read this book, unless yours are the same as ours.

1

The history of the sixth IIASA symposium on global modelling

The first book on global modelling to attract wide attention was *The Limits to Growth* (by D.H. Meadows *et al.*, Universe Books, New York, 1972). Because of the publicizing activities of the Club of Rome, and perhaps because the book addressed a deeply felt concern of mankind, *Limits* achieved a remarkable degree of visibility.

Scientists in fields where modelling was an accepted method—primarily economics, engineering, computer science, and management science—responded to this notoriety with varying degrees of applause, concern, hostility and amazement. Both advocates and critics agreed that future global modelling work should be subjected to the careful scrutiny of an international community of scholars. The only question to be resolved was how and where.

Seven months after publication of *The Limits to Growth,* the representatives of twelve national scientific organizations met in London to affix their signatures to the charter of a new and unique international scientific organization, the International Institute for Applied Systems Analysis (IIASA). The objectives of the Institute, as set forth in the charter, were threefold:

— to increase international collaboration, especially between scientists from East and West;
— to contribute to the advancement of science and systems analysis;
— to achieve application of systems analysis to problems of international importance.

While some people thought IIASA might undertake a leading role in the construction of a global model, the leadership of the Institute felt that a more valuable contribution would be possible if it did not become an advocate of a single approach in this rapidly evolving and highly controversial field. Instead it was decided that IIASA should focus on methodological issues, assuming the role of a 'sophisticated clearing house' for global modelling activities. As founding Director Howard Raiffa stated in a welcoming address to the first IIASA conference on global modelling:

> It is IIASA's intention to host, not a single symposium, but a series of symposia on global modelling efforts. In this respect, we are fulfilling our information-exchange agency role .... We would ourselves like to concentrate on methodological aspects of such modelling efforts. We hope we can play a constructive role in helping to identify critical methodological areas where more basic research work is needed. And we intend, in collaboration ... with others, to work on improving the methodology of such large-scale modelling efforts as this one. So we enter with the spirit of being a friendly critic.
>
> In no way whatsoever does IIASA ... think this modelling effort (Mesarovic/Pestel) is the best one available, nor do we as an institution necessarily endorse any of the conclusions of this modelling project .... We wish to strike a neutral pose in all these matters. We do assert, however, that we think that the problems addressed in these symposia are important ones for humanity, and we do hope that by such open and frank discussions of these methodological issues we collectively will become more sophisticated, more accurate, and more relevant.

Thus IIASA assumed, and has continued to play, an important monitoring role in the field of global modelling. Whenever a major model approached completion, IIASA organized an international conference devoted to presenting and discussing it. Through these conferences, members of the global modelling community have had the opportunity to meet, not only for formal discussions but also for the *Heurigen* (traditional Austrian wine-drinking parties) that are an important part of every major IIASA event. Thus, through IIASA's involvement, global modellers have become a 'community of scholars', not only in name but also in fact.

These models were discussed at the first five IIASA conferences on global modelling:

First Conference (April 1974):	The Mesarovic/Pestel model
Second Conference (October 1974):	The Bariloche model
Third Conference (September 1975):	The MOIRA model
Fourth Conference (September 1976):	The SARU model The MRI model (a Polish national model)
Fifth Conference (September 1977):	Input-output models, especially the UN world model and FUGI model

After the fifth global modelling conference, a consensus began to emerge that the time had come to attempt an appraisal of the state of the art. Seven different models had been completed and documented for scientific review. A community of model critics had emerged; while these critics had not made their own models, they were familiar with all of the efforts in the field and were interested in comparing and contrasting them. New global modelling teams were forming in the Wissenschaftszentrum, Berlin, and the USSR All-Union Institute for Systems Studies. Thus, it was felt strongly that the emphasis of the sixth conference should be on review, assessment, and constructive suggestions for the future, not on recriminations about the past.

Deciding to have such a conference was just the first step. In the beginning it was not at all clear how a meaningful assessment of such a diverse field could be achieved. The existing global models started from different paradigms, used different methods, and focused on different problems. There seemed to be no agreement among global modellers about answers to basic questions or even about what the basic questions were. It was obvious that the task of appraisal could not be accomplished through ordinary research, which would inevitably reflect the biases of the researchers. It had to include, in a very active mode of discourse, leading representatives of the major global modelling groups as well as scientists who had been involved in previous attempts to criticize and appraise the global models.

After a long period of discussion, both within IIASA and outside of it, an elaborate multistage process was designed and implemented. First, some twenty leading scholars and practitioners in the field of global modelling were asked to formulate questions they felt should be posed to the representatives of seven groups who had already completed global models. (Because of its national emphasis, it was decided not to include the MRI model.) More than two hundred proposed questions were received.

The responses were consolidated by a group of IIASA scientists into seven blocks of questions, focusing on the major issues of

- purposes and goals of global modelling,
- methodology,
- actors and policy variables,
- structural aspects of the models,
- testing the models,
- internal organization of global modelling projects,
- relations between modeller and user.

The questions were formulated in a future-oriented way, but were to be answered on the basis of experience gained with past and present modelling work.

Implementation of the plan to submit the questionnaire to representatives of the modelling groups proved to be more difficult than anticipated. Several groups had dissolved. Others did not have a definite spokesman. Furthermore, the replies defied easy comparison. Some modellers reported on their experiences and others did not. Some felt unable to answer the questions as they stood and reformulated them. Others provided lengthy and thoughtful statements of their general modelling philosophy. The responses are found in Chapter 4 of this volume.

At this point the aid of a number of 'evaluators' (critics and appraisers of the field) was enlisted. The original intention was to make this group completely model-independent, but this seemed, upon reflection, to be undesirable. Some of the better-known critics of the field (for example Peter Roberts and two of the co-editors of this volume) had also been closely associated with the development of a major model. Consequently, the organizers decided upon an 'Austrian solution': the evaluators would be model-independent except in a few cases where compelling reasons seemed to dictate otherwise. Each evaluator was asked to prepare, on the basis of the replies received, a paper that commented on one of the issues emphasized in the questionnaire. Their papers, along with the conference discussions, are found in Chapter 5.

The success of the conference was enhanced by the generosity of the *Stiftung Volkswagenwerk* (Volkswagen Foundation), which sponsored a preconference workshop for modelling-group representatives and evaluators. This workshop resolved some inconsistencies and allocated the limited time available for conference discussions.

The conference itself was organized around the seven main issues. In each session, the respective evaluator(s) first presented an introductory statement. This was followed by lively discussion. In the course of the conference, a few subtopics that proved to be of particular interest were set aside for a final period of 'general evaluation' in which a number of IIASA scientists working in different fields participated as active discussants. The last three sessions, as has become traditional, were devoted to brief reports on ongoing global modelling projects.

The departure of the conference participants from Vienna did not mark the end of the venture. In fact, it was only the midpoint. During the next month, each of the modelling groups was invited to revise its questionnaire responses, taking into account the discussions, the evaluators' comments, and the replies of other groups. Group representatives were also invited to contact the evaluators directly, if they felt that an evaluation had missed or misinterpreted an important point in their reply.

Commitments made in the euphoric last hours of a successful international conference are rarely fulfilled. However, to their credit, most of the groups submitted carefully reviewed, thoughtful revisions. Several also contacted evaluators directly. After their materials had been received at IIASA, they were sent to the evaluators, who were asked to revise their papers and comments. Again, a virtually complete set of responses was received. Thus a remarkable set of materials awaited the co-editors when they arrived at IIASA in mid-June 1979 to assemble the proceedings in final form. An intense two weeks was required for that effort. The resulting draft was circulated for comment during the ensuing year, the comments were used to guide a revision of the text. The final product of this process is now before you.

A short editorial appreciation

You may have been involved yourself in organizing international conferences and editing conference proceedings. If so, you will recognize that the process described here is unique in at least two respects.

The first is obvious.

The process that produced the material comprising this volume was a lengthy and complex one.

The second is obvious when you think about it.

The process succeeded only because of an extraordinary level of dedication on the part of a large number of people.

And we are not talking about people who are blessed with inordinate amounts of free time. For most of the modellers and evaluators seven-day weeks and twelve-hour days are a matter of routine.

You will find throughout this book ample evidence of commitment in the quality of the contributions. Unfortunately, you will not be able to share directly the sense of mutual respect and willingness to work cooperatively that characterized the conference discussions. The spirit of the conference is impossible to convey in words

(although Paul Neurath has come very close in his conclusion to Chapter 5, which you should be sure to read).

We hope that our own contribution to the process reveals to some degree the very deep admiration and appreciation we feel towards our colleagues in the community of global modelling.

Before beginning Chapter 2,
which is entitled

What are global models and what have they taught us about the world?

We would like to tell you why we wrote it.

It is written for a friend.

This is his story.

The lawyer's story

Under the linden trees outside IIASA's Schlossrestaurant, two global modellers of our acquaintance (whom we shall call D and J) were having lunch with a friend (whom we shall call H). H happens to be a lawyer, but he does not practise law.

H was explaining to D and J some of the problems he encounters in his job, which is to serve as liaison officer between IIASA and one of the national scientific organizations that sponsor the Institute.

Much of his time is devoted to answering questions about IIASA posed to him by politicians and bureaucrats. Much of the remainder is spent trying to help people at IIASA communicate with politicians and bureaucrats more effectively. Like many 'middle men', he often finds his task frustrating.

As the conversation progressed, D and J discovered that global modellers were a particular source of frustration to him.

'The main trouble with you global modellers,' H said, 'is that you have never made it clear to anyone what you do. I have enough trouble explaining any kind of mathematical modelling to politicians.'

' "What is it?" they want to know.
"Why do it?
Who does it?
What do you do it with?
Can I do it?
Will it help me to solve my problems?" '

'But *global* modelling,' he concluded,

'I can't even explain *that* to other *modellers!*'

2

What are global models and what have they taught us about the world?

Models

A model is any simplified, generalized image of reality. Maps, toy trucks, corporate balance sheets, statements such as 'united we stand, divided we fall', and equations such as $E = MC^2$ are all models. This paragraph is a verbal model of what models are.

A *good* model is not just any old simplification, but one that distills out of a complicated and confusing world the essence that allows one to achieve a particular purpose. A good model of an airplane, for the purpose of testing its flight characteristics in a wind tunnel, duplicates a real plane's shape and weight distribution, but it can be any color. A good model for a display in an airline company's window may have any weight distribution, may exaggerate the sleekness of the shape, but *must* be painted with the proper colors. In short, *A MODEL CAN BE MADE AND JUDGED ONLY WITH RESPECT TO A CLEAR PURPOSE.* The very best model is one that contains just what is needed for that purpose, and no more. It is *elegant,* which means, according to the dictionary, 'ingeniously simple'.

A basic assumption of mathematical modellers (and most philosophers and psychologists) is that the human mind operates entirely by creating models. Human decisions are virtually never based on accurate perceptions of all relevant properties of the world. They depend on biased and incomplete perceptions of whatever limited part of the world a single individual has experienced or has absorbed from other people's mental models of what they have experienced. Like the gentleman who is surprised to discover he has been speaking prose all his life, most people are only partly aware that they have been modelling all their lives. In every action you take— deciding whether to wear a raincoat today, what career to follow, how to price a product, whether to have a baby, run for political office, fight a war— you use abstracted pictures about how the world is and will be. For better or worse, human beings think with models and they cannot think without them.

Mental models can be ranged along several hierarchical levels, from the most temporary and shallow summaries of trivial recent events to the most deep, enduring descriptions of the basic nature of the world. The very deepest models are referred to in several places in this book as *paradigms*. Paradigms are unquestioned and nearly subconscious beliefs that serve to screen, order, and classify incoming impressions and thus to shape subsequent mental models. For instance, if one has a paradigmatic belief that humans are aggressive, selfish, and competitive, one sees and remembers greedy behaviour on the part of oneself and others, and one tends to forget or even *not to notice* altruistic, loving, cooperative behaviour. Paradigms are thus self-fulfilling hypotheses as long as they are at least partly congruent with the real world. Only when massive evidence contradicts them are they shifted or changed (for many examples, see Thomas Kuhn, *The Structure of Scientific Revolutions,* University of Chicago Press, 1964). A few of the paradigmatic assumptions that characterize the Western industrialized world view are

- that human beings are clearly separate from each other and that Man is distinct from and somewhat at war with Nature;
- that the rational, logical, sequential activities of the brain are superior to the non-rational, emotional, and intuitive activities;
- that human progress can be measured by material output and physical comfort;
- that work and leisure are different parts of one's life and that one must work in order to deserve leisure;
- that growth is good and bigger is better;
- that ultimately the world can be understood; that science is moving toward greater and greater understanding.

Since mental models on both the paradigmatic and the more superficial levels have sustained *homo sapiens* for hundreds of years, they must be fairly capable of reflecting the world and adapting to changes in the physical environment and in human purposes. Throughout the discussion of computer models that constitutes most of this book, the great strengths of mental models should not be forgotten. Most mental models are extremely subtle and elaborate. The human brain is an instrument that people would consider incredible and magnificent if it weren't the thing they consider with. The models you have about a person you know well, about a social system you have participated in, or about a field of abstract knowledge you have studied are amazingly sophisticated. Mental models can draw from a rich store of memories and perceptions, especially incisive perceptions about human quirks, motivations, behaviour, and values. Mental models are not constrained by academic disciplines or mathematical rigour. They can integrate widely varying types of information in strikingly creative combinations. They can balance incommensurables such as the aesthetic value of a tree left standing versus the economic value of a tree cut down. Mental models are often (not always) good at extracting the essential pieces of information relevant to a problem and ignoring the trivia. They are often elegant. Above all, mental models are quick and cheap and handy. They are used exclusively for at least 99.99 per cent of human decisions, and they probably should be.

Computer Models

So why make computer models?

The simplest answer is, because there are some problems that mental models cannot handle, such as:

- those that require a very precise answer (design of a nuclear power plant, calculation of a rocket trajectory);
- those that deal with interlocking systems of many components (scheduling oil-tanker routes, predicting the performance of a national economy);
- those that benefit from combining knowledge from many mental models or viewpoints (design of a long-term energy plan, assessing the environmental impact of a large dam);
- those that involve decisions where mistakes would be unacceptably costly or irreversible (simulating moon landings, calculating a sustainable harvest of whales, projecting whether carbon dioxide in the atmosphere will change the world's climate).

There is a general impression in the world that, as human societies evolve, problems like these are increasing— perhaps because mental models are solving the easier ones.

Given all these different kinds of problems that tax the capabilities of mental models, one would expect that there would be many different kinds of computer models, and there are. Some computer models are used directly to make on-line decisions about relatively straightforward but minutely complicated things, like the chemical operations in oil refineries, the order streams for multiproduct retailers, or the optimum mix for an animal feed. But very few problems are so mechanical and well understood that a computer can just take over the entire decision-making process. More usually the computer is used to calculate a few inputs to a policy that will ultimately be fashioned by the intricate give-and-take of mental models. A demographic model, for instance, may be used to calculate the characteristics of the work force of 1990, as an input to a labour policy. An econometric model may be used to forecast general economic growth, which will be one factor taken into account in an investment decision. A simulation model can tell a national leader what the economic repercussions of a strong or a weak energy conservation policy may be. Or a city model may serve to demonstrate clearly how the different city service functions (water, public transportation, solid

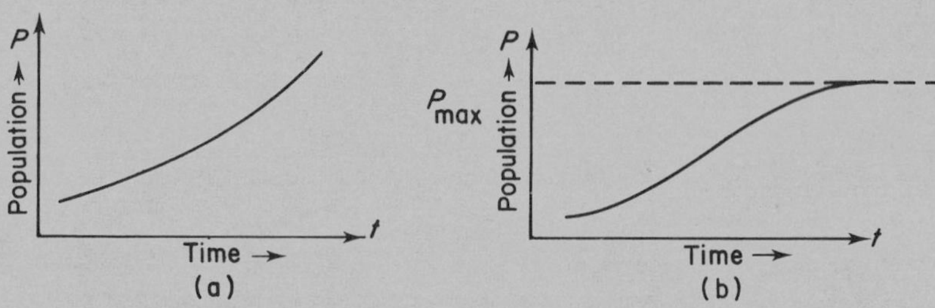

Figure 1. Models of population growth

waste removal) can be financed. In other words, computer models are *used* in many different ways, very few of which involve cranking out exact instructions about what someone should do. Mostly these models are used to keep track of many things, to run out forecasts under different sets of assumptions, or to provide a general conceptual framework that will help in thinking or communicating about a problem.

Mathematical models are nothing more than the expression of assumptions from mental models in the form of mathematical equations. If you think populations grow faster and faster over time, you can say this in words, or you can express it with the graph in Figure 1(a), or you can say it mathematically with the equation $\mathrm{d}P/\mathrm{d}t = rP$. If you think population levels off over time, as shown in Figure 1(b), then you can express this fact mathematically as $\mathrm{d}P/\mathrm{d}t = (P_{max} - P)rP$. The verbal statements, the time graphs, and the equations are just three different ways of saying the same thing, like 'no', 'nein', and 'nyet'. Also, like words in different languages, each form of expression is comprehensible only to people who have been trained to understand it. One of the main problems with mathematical models is that most people aren't trained to understand them, and so they must usually be accompanied by a translation back into ordinary words. The problem is compounded when the message is further translated from mathematics into the digital kinds of languages that computers can understand. Any process of translation requires a considerable amount of trust. The double-translation process from words to mathematics to computer code and back again makes people suspicious, and it should. It requires very good translators.

Mathematical models cannot contain any information that did not come from someone's mental model. They can *produce* information that is inherent in our assumptions but that we are not aware of because we have not pursued the logical implications of our assumptions. Here are a few very simple examples:

The assumption that	*implies mathematically that*
Electricity production will grow continuously at 7 per cent per year	The number (or size) of electrical generating plants will double in ten years and quadruple in twenty years
New inventory for a warehouse will be ordered by taking the difference between the desired inventory and the amount actually on hand	Inventory will never level off at the desired level, but will oscillate around a lower level than that desired
All parents in a given population want boy children and keep having children until they get a boy	The percentage of males in the population will be *lower* than it would be if there were not a preference for boys.*

At least theoretically, computer models can excell mental models in several ways:

- *Rigour.* The assumptions in computer models must be expressed completely and precisely; no ambiguities are possible. Every term must be defined, and assumptions must be mutually consistent. Computer modellers often mention that the discipline required to reformulate their mental models in mathematical terms is helpful in organizing and clarifying their thoughts even before any computer analysis takes place. Mental models are usually fraught with inconsistencies.
- *Accessibility.* Because all assumptions must be written out in order to communicate them to the computer, critics can examine, assess, and alter computer models. Mental models are unexaminable and uncriticizable; their assumptions are not even entirely clear to the modellers.
- *Comprehensiveness.* A computer model can contain much more information than any single mental model and can keep track of many more interrelations at one time.
- *Logic.* If programmed correctly, the computer can process even a very complicated set of assumptions to draw error-free conclusions. The human mind is quite likely to make errors in logic, especially if the logical chain is complex. Different people may agree completely about a set of assumptions and still fight about the conclusion to be drawn from them, if they have only mental models to rely on.
- *Flexibility.* Computer models can test a wide variety of conditions and policies, providing a form of social experimentation that is much less

*Compare Nathan Keyfitz, 'What mathematical demography tells that we would not know without it', Working paper No. 20, East-West Center, Honolulu, Hawaii, March 1972.

costly and time-consuming than tests in the real world. Different assumptions can be tested more easily and openly than they could be in a single mental model.

That is the sort of list that an enthusiastic computer modeller is likely to draw up to convince a skeptical client to employ him. The client, and in fact almost anyone outside the field of computer modelling, probably will feel a bit uncomfortable about all these claims, however, and for a very good reason. Those are *potential* advantages of computer models over mental ones, but they are by no means always realized. Computer models can become so supercomplex and be expressed in such esoteric jargon that they are less open to critical evaluation than mental models. They are not logically error free if they contain undetected errors in programming, typing, or interpretation. They are not more comprehensive than mental models if their mathematical properties require the omission of things that non-mathematical people know to be important. They cannot provide easy tests if each run requires several thousand dollars, several days of computer time, and six months of preprogramming and postprocessing. Computer modelling techniques are also based on deep, unexpressed, technical paradigmatic assumptions that are *not* congruent with mental models and often constrain what can be included. Problems of this sort are not hypothetical, as much of the rest of this book demonstrates.

So, on the one hand, there are human purposes, truly critical ones, for which mental models are inadequate and computer models offer hope. On the other hand, computer models have their own inadequacies and often fall far short of their potential. In this situation it is easy to resort to one or another extreme position: to reject the entire field of computer modelling and put one's faith in muddling through, or to dismiss the messiness of qualitative human thinking and try to computerize the world. The prime purpose of the Sixth Global Modelling Conference and of this book is to establish the grounds for a more knowledgeable position than either of these extremes. By drawing together the thoughts of the modellers and their critics, we hope to provide a description of the current strengths and weaknesses of computer models accurate enough to allow their makers to see ways of making them better, and their users to learn from the models what indeed can be learned. We hope everyone will remain skeptical of both mental and mathematical models, while at the same time using them, since they are, in fact, all we have.

Global Models

So much for computer modelling in general. Now what is *global* modelling?

GLOBAL MODELLING IS JUST COMPUTER MODELLING DONE TO

INVESTIGATE SOCIAL QUESTIONS OR PROBLEMS OF GLOBAL SCALE. For example, global models have addressed questions such as these:

- Is the exponential growth of the world's population and capital stocks sustainable by the world's resource base and ecosystems? What might happen if these resources become strained?
- How do national policies and international markets for trading food allow widespread hunger to exist? What kinds of policies could reduce hunger?
- Is it possible for all basic needs to be provided for the world's people over the next several decades? How?

Global modelling is distinguishable from other types of modelling of social systems only by the questions it asks. Its methods, strengths, and weaknesses are identical to those of all policy-oriented computer models. It draws from the same base of theory, data, and technique. Therefore, if there are any distinct properties of global modelling, they follow directly from the characteristics of global problems. Global models are sometimes (not always) more complex, more ambitious, and perhaps longer in time horizon than national or regional models. The global modelling groups have so far probably been more diverse in method, ideological background, disciplinary base, and team membership than most groups modelling a single nation or region. Global models have certainly been more in the public eye, and have aroused more interest and controversy than most other computer models. The reason for this is a matter of speculation; one hypothesis is that problems and conclusions of a global scale are naturally likely to arouse global reactions. No one can feel unaffected by pronouncements coming from a global model.

Critics of global modelling ask: 'Why do global modelling at all, or, having done some, why do more?' Apparently we global modellers have never answered the question very clearly, since it keeps being asked. Our lack of clarity may result from the fact that the answer seems so obvious to us. In our mental models the most important, urgent, compelling, undismissable problems are global. Nuclear armaments, depleting resources, growing strategic vulnerability of nations, abject poverty, threatened ecosystems are the problems we see, and we also see interconnections among all these problems. These are just the complex, interlocking, interdisciplinary sorts of issues that we know cannot be handled by mental models alone. We see no choice but to increase our understanding of them as much as possible with computer models, even though we're not sure how to make such models— even if we have to grope in the dark. Policies and technologies with long-term global implications are being suggested and implemented every day— expansion of nuclear energy generation, new international trade agreements, rising petroleum prices, dissemination of pesticides, SALT treaties, satellite communications. If experiments are being done on a global scale, it seems essential to have global models to test them, for the global system, above all, is not one we want to subject to costly mistakes or unforeseen, irreversible side effects.

What We May Have Learned So Far

The first global model was published in 1971. Depending on what one counts as a global model, about ten others have been begun since then. Seven of them have been brought to a stage of sufficient completion and documentation to be the focus of this book. Their stories are told in the next chapter. These models span a wide spectrum of purposes, methods, and paradigms. They have been based in the United States, Europe, South America, and Japan. Some have been made deliberately to refute others. Each one is a different and bold experiment and each has obvious weaknesses, most of which the modellers themselves are aware of and discuss at length in this book.

Given the newness of the field and the variety of approaches tried so far, we did not expect before the conference that there would be much consensus, even about the state of the art. We certainly did not expect to be able to extract substantive statements about the world that could be agreed upon by all modellers. In going over the conference proceedings, however, we were surprised to find that both expectations were wrong. Methodological disagreements remain, but some common suggestions emerged about how global modelling has been done so far and how it might be improved. These are contained in Chapter 6.

And even more surprising, given the very different backgrounds of the models and modellers, the qualitative conclusions drawn by all of them about the current state of the world and its possible futures are quite similar. This does not imply by any means that the modellers agree on everything or that their precise numerical forecasts coincide. But on the broadest level, when one gets beyond the differences in reporting styles, levels of aggregation, and specific numbers, one finds some consistent messages. They are, as one might expect, general messages about the way global systems operate, rather than detailed policy advice (much of which did come out of the models, but in such diverse and non-commensurate forms that areas of agreement are difficult to find). In one sense the messages are not surprising; all of us know them at some level. Yet in another sense, they are revolutionary; if everyone internalized them deeply and acted on them, the world would be a different place.

The areas of agreement are set forth on the ensuing blue pages.

Each statement in this list meets, we believe, two criteria:

- All global modellers would agree to it.
- At least one group of global modellers considers it very important.

1. There is no known physical or technical reason why basic needs cannot be supplied for all the world's people into the foreseeable future. These needs are not being met now because of social and political structures, values, norms, and world views, not because of absolute physical scarcities.
2. Population and physical (material) capital cannot grow forever on a finite planet.
3. There is no reliable and complete information about the degree to which the earth's physical environment can absorb and meet the needs of further growth in population and capital. There is a great deal of partial information, which optimists read optimistically and pessimists read pessimistically.
4. Continuing 'business-as-usual' policies through the next few decades will not lead to a desirable future— or even to meeting basic human needs. It will result in an increasing gap between the rich and the poor, problems with resource availability and environmental destruction, and worsening economic conditions for most people.
5. Because of these difficulties, continuing current trends is not a likely future course. Over the next three decades the world's socioeconomic system will be in a period of transition to some state that will be, not only quantitatively, but also qualitatively, different from the present.
6. The exact nature of this future state, and whether it will be better or worse than the present, is not predetermined, but is a function of decisions and changes being made now.
7. Owing to the momentum inherent in the world's physical and social processes, policy changes made soon are likely to have more impact with less effort than the same set of changes made later. By the time a problem is obvious to everyone, it is often too late to solve it.
8. Although technical changes are expected and needed, no set of purely technical changes tested in any of the models was sufficient in itself to bring about a desirable future. Restructuring social, economic, and political systems was much more effective.
9. The interdependencies among peoples and nations over time and space are greater than commonly imagined. Actions taken at one time and on one part of the globe have far-reaching consequences that are impossible to predict intuitively, and probably also impossible to predict (totally, precisely, maybe at all) with computer models.
10. Because of these interdependencies, single, simple measures intended to reach narrowly defined goals are likely to be counterproductive. Decisions should be made within the broadest possible context, across space, time, and areas of knowledge.

11. Cooperative approaches to achieving individual or national goals often turn out to be more beneficial in the long run to all parties than competitive approaches.
12. Many plans, programs, and agreements, particularly complex international ones, are based upon assumptions about the world that are either mutually inconsistent or inconsistent with physical reality. Much time and effort is spent designing and debating policies that are, in fact, simply impossible.

In short: CHANGE IN THE STATUS QUO IS CERTAIN. IMPROVEMENT IN THE STATE OF THE WORLD IS BY NO MEANS IMPOSSIBLE AND BY NO MEANS GUARANTEED. WE ARE A LONG WAY FROM KNOWING EVERYTHING WE NEED TO KNOW. AND YET WE KNOW ENOUGH ABOUT WHERE WE WANT TO GO AND HOW TO GET THERE TO GET STARTED. THE SITUATION IS NOT HOPELESS: IT IS CHALLENGING.

Another way to think

Looking over the foregoing list,
We are struck by a contradiction.

The global models seem to say

FIRST, that the world is not in very good shape and is heading in an even worse direction, although there are some better directions possible.
In other words, WE SHOULD DO SOMETHING.

SECOND, that the world is very hard, if not impossible, to figure out, even for the experts with their big-gun computers.
In other words, WE CAN'T BE SURE WHAT WE SHOULD DO.

No wonder people are frustrated by global models!

If you like to think that the world is capably controlled
By people who know what they are doing
(one of whom might be you)
And that tomorrow will be much like yesterday,

Messages from global models must be disconcerting.

But there are other ways to think.

One could think that

What we're doing now isn't working.	But there are many other things to do.
We can't be experts.	But we can be experimenters.

(No one likes to be the subject of an experiment,
but let's face it,
we already are.)

Instead of pretending that we're sure and that we know what is right, and instead of moving heaven and earth to keep those present trends continuing,

We could keep our eyes and ears and minds open
be honest about how uncertain we are
try new things on a small, safe scale
learn as much as possible from all sources,
even global models
and regard the whole situation not as a crisis
but as an adventure.

3

*The cast of characters: an introduction to global models**

Questionnaire responses from respresentatives of the seven modelling groups formed the background material on which the conference was based. These responses comprise an unusual store of insight and experience about the craft of modelling. They become more valuable, however, if one knows something about the origins, objectives, structures, and major conclusions of the modelling projects.

Accordingly, in this section we provide a brief description of each project and its principal products. Some comparisons and contrasts of the models, emphasizing important distinguishing characteristics, are included.

Table 1 summarizes several important features of the seven projects, including

- the name of the project directors and core team members,
- the academic disciplines of the core team members,
- the modelling paradigm that was most influential in the development of the model,
- the primary publications produced by the project,
- where the team was located when it did most of its work,
- the date the model was presented at IIASA.

Most of the terms used in the table are self-explanatory, but two terms, 'modelling paradigm' and 'core' groups should be discussed briefly.

Paradigms were described in Chapter 2 as the unquestioned and nearly subconscious beliefs that serve to screen, order, and classify incoming impressions. In *The Structure of Scientific Revolutions* (University of Chicago

*Some of the material in this chapter builds upon two review articles on global modelling by John Richardson that appeared in *Futures,* 10:5 (October 1978) pp. 388-404 and *Futures,* 10:6 (December 1978) pp. 476-91.

Table 1. The major global models presented at IIASA conferences[a]

Project directors (*) and core team members	Disciplines	Type of model (paradigm)
The Forrester /Meadows models		
Jay W. Forrester (*)	Engineering/management	System dynamics
Dennis W. Meadows (*)	Management science	
William W. Behrens	Engineering	
Donella H. Meadows	Biophysics	
Roger Naill	Physics	
Jørgen Randers	Physics	
The Mesarovic/Pestel model		
Mihailo D. Mesarovic (*)	Engineering	Eclectic within a multilevel framework; includes input-output, system dynamics, and econometric components
Eduard Pestel (*)	Engineering	
W. Peter Clapham	Biology/ecology	
Barry B. Hughes	Political science	
John M. Richardson, Jr.	Political science	
Thomas Shook	Computer engineering	
The Bariloche model (Latin American world model)		
Amilcar Herrera (*)	Ecology	Optimization
Graciela Chichilnisky	Mathematics	
Gilberto Gallopin	Ecology	
Gilda de Romero Brest	Education	
Hugo Scolnik	Mathematical economics	
Carlos Mallman	Mathematics	
The MOIRA model		
Hans Linnemann (*)	Economics	Econometrics/optimization
R.J. Brolsma	Economics	
J.N. Bruinsma	Economics	
P. Buringh	Soil science	
H.D.J. van Heemst	Agronomics	
J. de Hoogh	Agr. economics	
M.A. Keyzer	Economics	
G.J. Staring	Soil science	
C.T. de Wit	Ecology	
The SARU model		
Peter Roberts (*)	Physics	Classical economics (with a dash of system dynamics)
M.R. Goodfellow	Physics	
J.B. Jones	Physics	
David Norse	Biology	
K.T. Parker	Control engineering	
W.G.P. Phillips	Engineering and economics	
The FUGI model		
Yoichi Kaya (*)	Engineering	Highly eclectic, incorporating input-output, optimization, and systems dynamics approaches
M. Ishikawa	Engineering	
H. Ishitani	Engineering	
A. Onishi	Economics	
K. Shoji	Engineering	
Y. Suzuki	Engineering	
T. Yamamoto	Economics	
The United Nations world model		
Wassily Leontief (*)	Economics	Input-output
Anne Carter	Economics	
Peter Petri	Economics	

[a]As noted above, the Forrester Meadows models predated IIASA, but are nonetheless included in this book.

Location of the project	Date presented at IIASA	Principal publications
Sloan School of Management, MIT, Cambridge (Mass.), USA	– –	*World Dynamics,* Wright Allen Press, Cambridge (Mass.), 1971. *The Limits to Growth,* Universe Books, New York, 1972. *Dynamics of Growth in a Finite World,* Wright Allen Press, Cambridge (Mass.), 1974.
Systems Research Center, Case Western University, Cleveland (Ohio), USA/Technical University, Hannover, Fed. Rep. of Germany	29 Apr.- 3 May 1974	*Mankind at the Turning Point,* Dutton, New York, 1974.
Fundaçion Bariloche, San Carlos de Bariloche, Rio Negro, Argentina	7-10 Oct. 1974	*Catastrophe or New Society,* International Development Research Centre, Ottawa, 1976
Economic and Social Institute, Free University, Amsterdam, Netherlands, and Agricultural University, Wageningen, Netherlands.	22-25 Oct. 1975	*MOIRA — Model of International Relations in Agriculture,* North-Holland Publishing Co.,
Systems Analysis Research Unit, Department of the Environment, London, UK	20-23 Sept. 1976	*SARUM 76 — Global Modelling Project,* Research Report No. 19, UK Department of Environment and Transport, London, 1977. *SARUM and MRI,* IIASA Proceedings Series, Vol. 2, Pergamon Press, 1978.
Engineering Research Institute, Faculty of Engineering, Tokyo University, Tokyo, Japan	26-29 Sept. 1977	*Future of Global Interdependence,* Report, presented at the Fifth IIASA Global Modelling Conference, 1977. *FUGI,* IIASA Conference Series (in print).
Economic Research Center, New York University Brandeis University	26-29 Sept. 1977	*The Future of the World Economy,* Oxford University Press, New York, 1977.

Press, 1964), Thomas Kuhn has observed that a major part of the training of scientific disciplines is, in effect, a process of socialization to a particular world view. Included in this view is a host of subconscious and unrecognized values and assumptions about the nature of the world. For example, the systems dynamics modeller sees the world as a collection of feedback loops and non-linear relations. He or she values the kind of understanding that comes from clarifying relations between causal structure and dynamic behaviour. An econometrician sees the world through the prism of economic theory and values precise predictions of the near future. It is difficult for either practitioner to achieve a sufficient degree of detachment to explain his or her viewpoint to the other. We have used our own judgement in associating paradigms with modelling projects—others might classify them somewhat differently.

The *core team* of a modelling project is a small subset, usually no more than five or six individuals, of the overall project team. In some of the projects described below more than fifty collaborators have been identified, but such listings communicate little information about who actually shaped the project and did most of the work. The core team worked closely, if not always harmoniously, throughout most of the life of the project and made most of the major decisions about the model. They typically had a deep personal commitment to what they were doing. This made possible real sacrifices and a level of productivity that would have been inconceivable under other circumstances. The identification of core team members is also based upon the judgement of the editors. A more detailed discussion of the internal organizations of global modelling project teams will be found in the rest of this chapter.

The Forrester/Meadows Models

For many individuals the World3 model described in *The Limits to Growth* (Meadows *et al.,* 1972) is the only global model. All too often, references to *Limits* evoke images of a gloomy future in which the most outstanding features are 'no growth' and 'collapse in the mid-twenty-first century'. However, this is not an accurate reflection of the model results. The complexities of computer models, even relatively simple ones, do not lend themselves to simple, sensational contemporary reporting in the media. The World3 modellers learned this the hard way, and later global modellers heeded the lesson.

In *World Dynamics,* the first publication on global modelling, which appeared in 1971, Jay Forrester has described the events that led up to the global modelling project at the Massachusetts Institute of Technology:

> On June 29, 1970, I attended a meeting of the Club of Rome in Bern, Switzerland. The Club of Rome is a private group numbering some 75 members

> from many countries who have joined together to find ways to understand better the changes now occurring in the world.* The members act as private citizens. They are not in governmental decision-making positions. Their orientation is activist—that is, they wish to do more than study and understand. They wish to clarify the course of human events in a way that can be transmitted to governments and peoples to influence the trends of rising population, increasing pollution, greater crowding, and growing social strife.
>
> At the time of the Bern meeting, The Club of Rome had already planned a project on 'The Predicament of Mankind'. Preliminary goals, a survey of methodologies and a statement of the 'problematique' had been prepared by Aurelio Peccei, Eduard Pestel, Alexander King, Hasan Ozbekhan, Hugo Thiemann, and others. The objective of the project was to understand the options available to mankind as societies enter the transition from growth to equilibrium ....
>
> The June meeting was held to review the status of the project, which was about to begin. Discussion in the meeting revealed that a suitable methodology had not been identified to deal with the broad sweep of human affairs and the ways in which major elements of the world ecology interact with each other. Because the 'system dynamics' approach as already developed at MIT seemed well suited, the group was invited to Cambridge to determine firsthand if they agreed that the methods then existing would be suitable for the next step in the project. As a result, a meeting convened on July 20 for ten days of study, presentations and discussion. (Forrester, 1971, pp. vii, viii)

In the short period of time between the June and July meetings, Forrester produced and documented a simple global model (sometimes called World1), which served as a vehicle for the presentations and discussions, and which was revised into World2, the subject of *World Dynamics*. When, shortly thereafter, a decision was made to go forward with the project, World2 became the take-off point for development of the more complex World3 or *The Limits to Growth* model.

The MIT group, headed by Dennis Meadows, was the youngest, and among the most visibly enthusiastic of the global modelling teams. With encouragement from the Executive Committee of the Club of Rome, the team produced an elaborated model and a popular description of it in less than two years. Under the auspices of the Club of Rome, Meadows and his colleagues presented their findings to high-level groups throughout the world. *The Limits to Growth* sold more than three million copies and was translated into twenty-three different languages (see principal references below).

World3 is one of the most criticized models of all time. Initially, much criticism focused on the fact that the team, with strong urging from the Club of Rome, elected to publish a popularized version of their work before the details had been subjected to scrutiny by the scientific community. This situation was quickly rectified, however. The DYNAMO computer language is virtually self-documenting, and Jay Forrester has always insisted on the

*The membership of the Club of Rome now numbers close to a hundred. Forrester's portrayal is still a good characterization of the Club and its activities. (Incidentally, Jay Forrester, Dennis Meadows, Mihailo Mesarovic, and Gerhart Bruckmann are all members of the Club of Rome).

highest standards of documentation from his students. The MIT team made their equations, in documented form, readily available, and the model was run in computer centres around the world long before its technical documentation was formally published. Today, thousands of graduate students and hundreds of their professors have become familiar with the technical details of World3. With the publication of the *Dynamics of Growth in a Finite World* (Meadows *et al.*, 1974), World3 became one of the most fully documented of the global models.

Those who believed that continued growth was essential to the preservation of Western industrial society and to the betterment of mankind attacked the principal conclusions. Scientists and political leaders from the Third World argued that the model was essentially an ideological statement from the developed world. (See the discussion of the Bariloche model, which follows on page 44). Economists attacked the world view of systems dynamics, which differed fundamentally from their own.

Much of the strident criticism of *The Limits to Growth* and World3 has now subsided. This is due in part to the fact that concrete indicators of some of the predicted crises have now become widely apparent. For example, it is hard to deny that there is a global energy problem, however optimistic one may feel about the prospects for solving it. Also, economists and others have gained a greater sense of humility towards global modelling from their own attempts to do it. The debate has not ended, as this book attests, but it has become more constructive and more collegial.

Purpose of the model

In Chapter 2, we emphasized that A MODEL CAN ONLY BE EVALUATED MEANINGFULLY WITH RESPECT TO A DEFINED PURPOSE. It is our impression that much of the discussion on World3 failed to take into account the purposes of the model and the decisions about model structure that followed therefrom.

The purposes that motivated the project were described by the Executive Committee of the Club of Rome in their preface to *The Limits to Growth* (1972):

> One was to gain insights into the limits of our world system and the constraints it puts on human numbers and activity ... a second objective was to help identify and study the dominant elements, and their interactions that influence the long-term behavior of world systems.

The statement of objectives by the Meadows team is somewhat more technical (Meadows *et al.*, 1974, p.8):

> Figure 1-1 illustrates the four possible behavior modes that a growing population can exhibit over time. The mode actually observed in any specific case will depend

on the characteristics of the carrying capacity—the level of population that could be sustained indefinitely by the prevailing physical, political, and biological systems—and on the nature of the growth process itself. One of these basic behavior modes must characterize any physically growing quantity, such as pollution, productive capital, or food output. *The purpose of World3 is to determine which of the behavior modes shown in Figure 1-1 is most characteristic of the globe's population and material outputs under different conditions and to identify the future policies that may lead to a stable rather than an unstable behavior mode.*

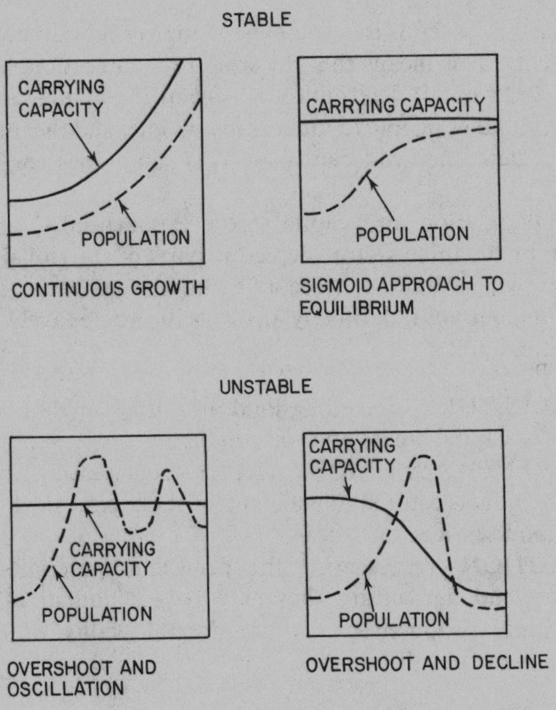

Figure 1-1 Four possible modes of population growth (Reprinted by permission of the MIT Press)

Given this ecological view of the global system, and a specific motivating concern with limits, the broad conclusions that emerged from the model are, perhaps, not surprising. The probability that limits exist and would be reached sometime was part of the initial problem definition (the most important model assumptions often occur in the problem definition).

Structure of the model

Figure 2 shows a simplified picture of the most important interrelations in the World3 model. The boxes represent the five major *sectors* (or submodels) of the overall model.

Two of the sectors focus on things that grow in the global system:

1. POPULATION– incorporating the effects of all economic and environmental factors that influence the human birth and death rate, and thus population size.
2. CAPITAL– including the manufactured means of producing industrial, service, and agricultural outputs, their growth through investment and decline through depreciation

Growth in these sectors tends to occur exponentially (like compound interest in a savings account). This means that the size of the important variable in each sector (people and capital) determines the size of the annual addition to that sector. In short, it takes people to make more people, and the more people you have, the more new ones you can make (the same goes for machines and factories).

In World3, population and capital grow exponentially, unless they are impinged upon by the three sectors depicting parts of the global system whose capacity for growth is limited. (According to the authors, technology can expand the limits, but not indefinitely and only at progressively greater costs.) These three sectors are:

1. AGRICULTURE– including land and other factors influencing the effects of capital inputs on food production.
2. NON-RENEWABLE RESOURCES– representing the fuel and mineral inputs required to make use of the capital stock for producing goods and services.
3. POLLUTION– representing the persistent materials produced by industry and agriculture that may reduce human life expectancy, agricultural productivity, or the normal ability of ecosystems to absorb harmful substances.

Figure 2 also shows some of the key variables that affect and are affected by what goes on in other sectors. For example, the effectiveness of birth control is partially determined by the overall state of development of the economy, the fertility of land is partially determined by the presence of persistent pollution, and so forth.

The most important assumptions in this model can be summarized roughly in words (all of the relations are highly non-linear):

1. Population increase (the difference between the birth rate and the death rate) is influenced by crowding, food intake, pollution, and the material standard of living. A rise in any of these four factors tends to drive the birth rate downwards. The death rate decreases with increasing food intake and the material standard of living, and increases with increasing pollution and (to a much lesser extent) crowding.

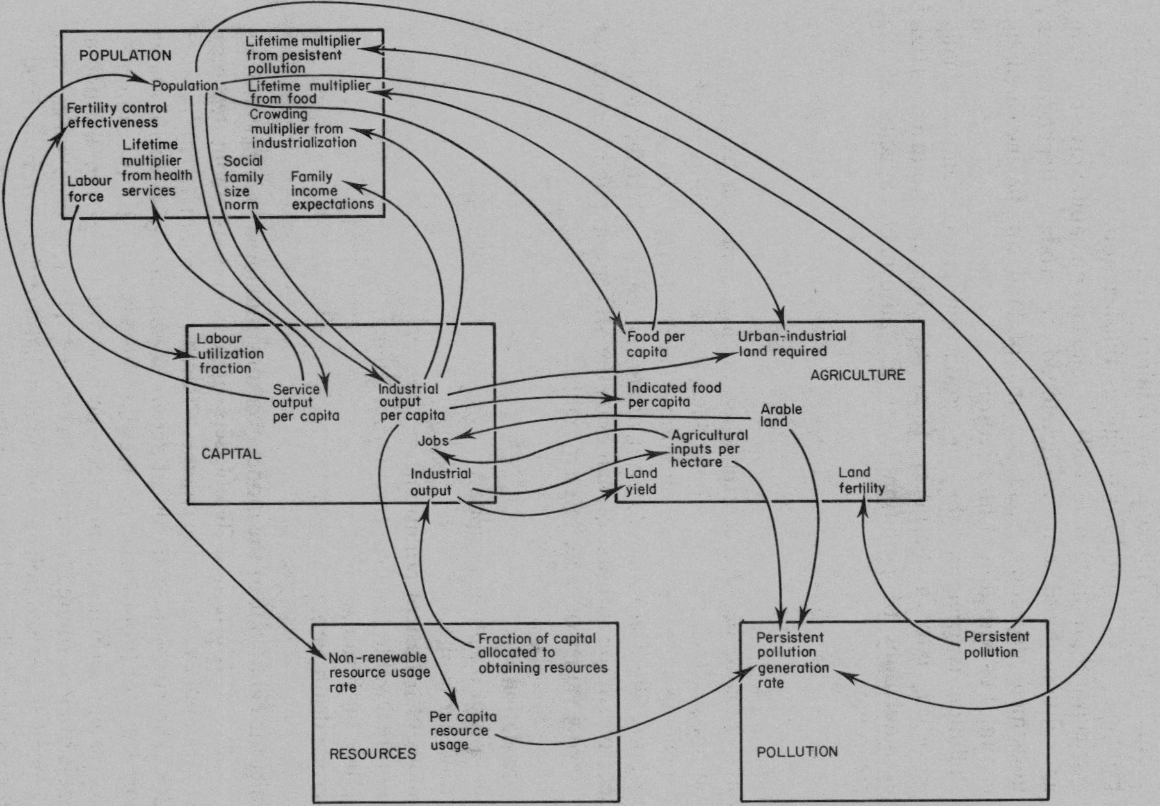

Figure 2. Interactions among the five basic sectors of World3 (from Meadows *et al.*, 1974, p. 11. Reprinted by permission of the MIT Press)

2. The material standard of living depends on the level of capital (relative to the size of the population) and the productivity of capital.
3. Non-renewable resources are continually used up by the production process. The lower the level of nonrenewable resources, the more capital must be allocated to obtaining resources, and thus the productivity of capital for producing finished goods is less.
4. Agricultural production depends on land and on capital investment in agriculture. Land can be developed or eroded, depending on investment decisions. Yield per unit of land can be increased by capital, but with diminishing returns.
5. Pollution is generated by the production process and gradually absorbed into a harmless form by the environment. High accumulations of pollution lower the absorbing capacity of the environment.

Results

One always needs to be very careful when talking about the 'results' of a simulation model. Such models will produce, literally, an infinite number of results, depending on the policy assumptions. The results reported especially in a popular book like *The Limits to Growth* are chosen to reflect the concerns and perspectives of the authors. In their book the MIT group published output graphs showing values of these seven variables:

1. population,
2. industrial output per capita,
3. food per capita,
4. index of persistent pollution,
5. non-renewable resource fraction remaining,
6. crude birth rate,
7. crude death rate.

They depict the behaviour of the model over a two-hundred-year period from 1900 to 2100 for fourteen different sets of assumptions about technical progress, social policy, and value changes— that is fourteen different 'runs' of the model. The supporting technical report, *Dynamics of Growth,* shows many more runs for more variables. We show here two illustrative examples.

The first example (Figure 3) provides the basis for the 'gloom and doom' image that is often attributed to the World3 model. This reference run assumes that the general values and policies that guided that world system from 1900 to 1970 will continue unchanged into the future.

In this graph, the portion of greatest interest lies between the years 2000 and 2050. During this period, both population and industrial output per capita grow 'beyond sustainable levels' and decline. Depletion of non-renewable resources, leading to a generally depressed economy, is the principal cause of

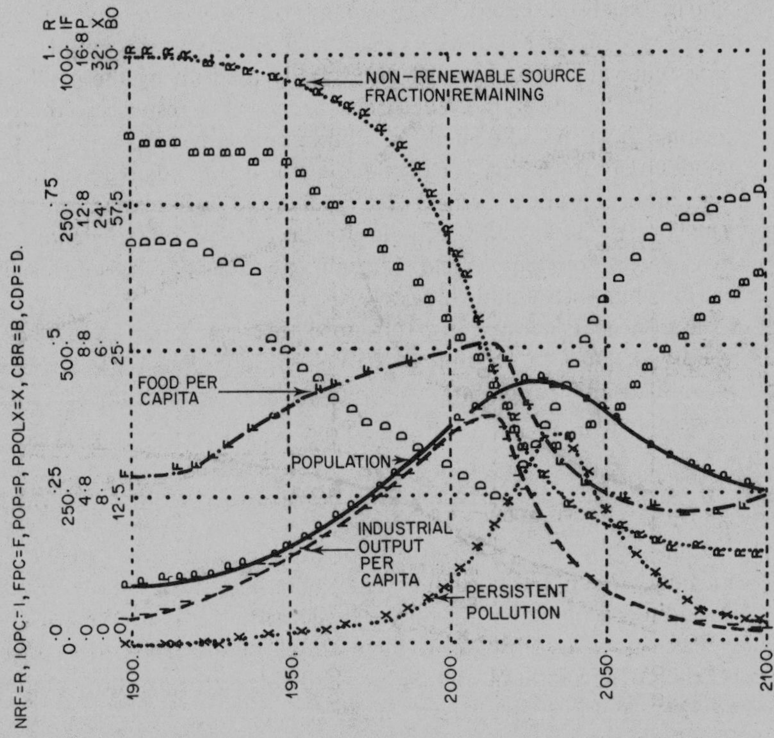

Figure 3. The World3 reference run (*The Limits to Growth: A report for the Club of Rome's Project on the Predicament of Mankind,* by Donella H. Meadows, Dennis L. Meadows, Jorgen Randers, William W; Behrens III. A Potomac Associates book published by Universe Books, N.Y., 1972. Graphics by Potomac Associates)

this decline. (For the reasons why the model system generates this 'overshoot and collapse', see pages 125-126 in Chapter 4).

Because global models are not intended to predict the future but to change it, 'results' such as depicted in Figure 3 only begin the process of using the model. The objective is to use one's understanding of the global system to identify strategies that can lead to more favourable futures. (What is meant by 'favourable' is, of course, a value judgement.)

Figure 4 shows an 'equilibrium through adaptive policies', a type of future favoured by the MIT group. Comparison of this picture with Figure 3 reveals a very different situation following the year 2000. The precipitous declines in population and industrial output are nowhere in evidence. Instead, growth is slower during the years preceding the turn of the century and then things stabilize at a much higher level. Population reaches about 6 billion and holds steady. Pollution first increases and then declines. The average world-wide standard of living is at about the current European level. According to the

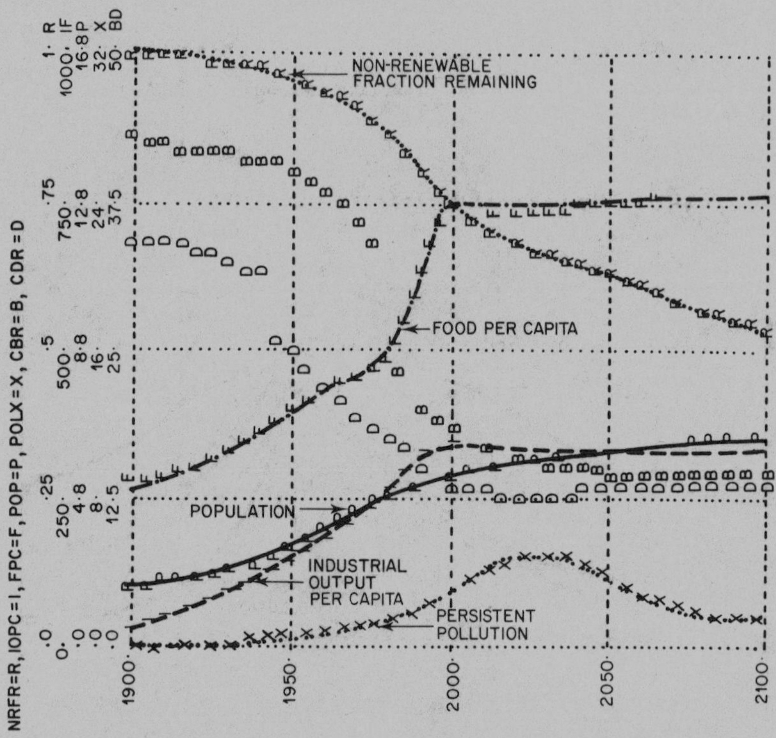

Figure 4. An equilibrium through adaptive policies. (*The Limits to Growth: A report for the Club of Rome's Project on the Predicament of Mankind,* by Donella H. Meadows, Dennis L. Meadows, Jorgen Randers, William W. Behrens III. A Potomac Associates book published by Universe Books, N.Y., 1972. Graphics by Potomac Associates)

modellers, this scenario could follow the implementation of 'adaptive technological policies that increase resource recycling, reduce persistent pollution generation and increase land yields, combined with social policies that stabilize population and industrial output'. Other runs show that speedy implementation of such policies will be most effective and that delays or halfway measures can be quite costly.

Numerous runs like these provided the basis for the very widely reported central conclusions of *The Limits to Growth:*

1. If the present growth trends in world population, industrialization, pollution, food production, and resource depletion continue unchanged, the limits to growth on this planet will be reached sometime within the next one hundred years. The most probable result will be a rather sudden and uncontrollable decline in both population and industrial capacity.

2. It is possible to alter these growth trends and to establish a condition of economic stability that is sustainable far into the future. The state of global equilibrium can be designed so that the basic material needs of each person on earth are satisfied and each person has an equal opportunity to realize his individual human potential.
3. If the world's people decide to strive for this second outcome rather than the first, the sooner they begin working to attain it, the greater their chances of success will be.

Later developments

Although the World3 model is still in use extensively, especially for educational purposes, none of the members of the Forrester/Meadows group is still engaged in global modelling. The group feels that it made the models for a specific purpose, this purpose has been fulfilled, and new purposes demand new models. Forrester at MIT is building an elaborate model of the US national economy. The Meadows group at Dartmouth is making models of the US energy supply system and regional models for managing renewable resources. Jørgen Randers has established a group making similar models for Norway. Roger Naill heads the Office of Analytic Studies of the US Department of Energy. Bill Behrens is an organic farmer in Maine. Each would describe his or her current activity as consistent with the message of *The Limits to Growth* and a further step towards understanding how to design a sustainable social system that is not based on the ideal of growth.

'And so, extrapolating from the best figures available, we see that current trends, unless dramatically reversed, will inevitably lead to a situation in which the sky will fall.' (From *The New Yorker*)

The principal references

Forrester, J.W. (1971). *World Dynamics,* Wright-Allen Press, Cambridge, Mass. (technical documentation of World2). Second edition 1973 (with new material on physical and social limits).

German: *Der Teuflische Regelkreis,* Deutsche Verlags-Anstalt, Stuttgart, 1971.

Italian: *Dinamiche Mondiali,* Etas Libri, Milan, 1974.

Japanese: *World Dynamics,* Nippon Keiei Shuppankei, Tokyo, 1971.

Meadows, Donella H., Meadows, Dennis L., Randers, Jørgen, and Behrens, William W. III (1972).

The Limits to Growth, Universe Books, A Potomac Associates Book, New York (popular account of World3). Reprinted in English by Signet Books, New York, 1974.

Brazilian: *Limites do Crescimento,* Editora Perspectiva S.A., Av. Brigadeiro Luis Antonio, 3025 Sao Paulo, Brazil, 1974.

Danish: *Graenser for Vaekst,* Gylendal Nordisk Forlag, Klareboderne 3, 1115 Copenhagen K, Denmark, 1972.

Dutch: *Rapport van de Club van Rome,* Uitgeverij Het Spectrum NV, Park Voorm 4 De Meern, Postbus 2073 Utrecht, Holland, 1972.

English: *The Limits to Growth,* Pan Books Ltd., 33 Tothill St., London SW1, England, 1971.

Finnish: *Kasvun Rajat,* Kustannusosakeyhtio Tammi, Helsinki, Finland, 1974.

French: *Halte A la Croissance?,* Librairie Artheme Rayard, 6 Rue Casimir-Delavigne, Paris 6[e], France, 1972.

German: *Die Grenzen Des Wachstrums,* Deutsche Verlags-Anstalt, Postfach 209, Neckarstrasse 121, 7 Stuttgart 1, Germany, 1972.

Icelandic: *Endimork Vaxtarins,* Hid Islenzka Pjodvinafelag, Skalholtsstig 7, Reyjavik, 1974.

Italian: *I Limiti Dello Sviluppo,* Mondadori Publishing Co., 20 Via Vianca Di Saveoia, 20122 Milano, Italy, 1972.

Japanese: *Seicho No Genkai,* The Diamond Publishing Co., Diamond Bldg. 1-4-2, Kasumigaseki, Chiyoda-ku, Tokyo, Japan, 1972.

Norwegian: *Hvor Gar Grensen.* Capallens Forlag, Kirkegatan 15, Oslo, Norway, 1972.

Persian: *Mahsosoiathai Roshd,* Anjoman Meli Hegazat Manabe Tabie Va Mohit Ensani, Teheran, Iran, 1974.

Polish: *Granice Wzrostu,* Panstwowe Wydawnictwo Ekonomiczne, Warsaw 1, Poland, 1973.

Portuguese: *Os Limites do Crescimento,* Publicacoes Com Quizote, Rua Luciano Cordeiro, 119, Lisbon 1, Portugal, 1974.

Slovene: *Meje Rasti,* Cankarjeva Zalozba, Ljubljana, Yugoslavia, 1974.

Spanish (Mexico): *Los Limites de Crecimiento,* Fondo de Cultura Economica, Av. Universidad, 975, Mexico 12, D.F., 1972.

Swedish: *Tillvaxtens Granser,* Albert Bonniers Forlan, 9A Kammadaregatan, Stockholm, Sweden, 1972.

Vietnamese: *Gioi Han Phat Trien,* Hien-Dai Thu-xa 28, Phung-Khac-Khoan, Saigon, 1974.

Yugoslavian: *Granice Rasta,* Stvarnost, Rooseveltov trg 4, Zagreb, 1974. (Also editions in Korean, Arabic, Hindi, Punjabi)

Meadows, Dennis L., and Meadows, Donella H. (Eds.) (1973). *Toward Global Equilibrium:*

Collected Papers, MIT Press, Cambridge, Mass. (technical documentation of

several submodels and discussion of special concepts in World3).
German: *Das Globale Gleichgewicht,* Deutsche Verlags-Anstalt, Stuttgart, 1974.
Italian: *Verso un Equilibrio Globale,* Arnoldo Mondadori, Milan, 1973.

Meadows, Dennis L., Behrens, William W. III, Meadows, Donella H., Naill, Roger F., Randers, Jørgen, and Zahn, Erich K.O. (1974). *Dynamics of Growth in a Finite World,* MIT Press, Cambridge, Mass. (technical documentation of World3).
French: *Dynamique de la Croissance dans un Monde Fini,* Economica, 49 Rue Hericart, 75015 Paris, 1977.

Meadows, Dennis L. (Ed.) (1977). *Alternatives to Growth-I: A Search for Sustainable Futures,* Ballinger Publishing Co., Cambridge, Mass. (collection of non-technical papers on the steady-state society).

Principal critiques

Cole, H.S.D., Freeman, C., Jahoda, M., and Pavitt, K.L.R. (1973). *Models of Doom,* Universe Books, New York (technical critique of World3).

Non-technical critiques and discussions of *The Limits to Growth* number in the hundreds and cannot all be listed here. For a partial listing, see *Growth and Its Implications for the Future* (4 vols), Serial Nos. 93-7, 93-28, 93-29, 93-35, Committee on Merchant Marines and Fisheries, The 93rd Congress, US Government Printing Office, Washington, D.C., 1973, 1974.

The Mesarovic/Pestel Model

The origins of what came to be known as the Mesarovic/Pestel model can be traced to a seminar organized by Dennis Meadows and held at MIT during the fall of 1972. (In some European circles, the model is referred to as the 'Pestel/Mesarovic model'. The correct name for the early versions, 'multilevel hierarchical regionalized world model' may suggest why a more cryptic label was often used. 'World integrated model' (WIM) is the name used to describe the latest versions, which are largely the work of Mesarovic's group in Cleveland, Ohio.) The subject of the seminar was the *Theory of Hierarchical Multilevel Systems* (Mesarovic *et al.,* 1970) developed by Mesarovic and his colleagues at the Case Western Reserve University's Systems Research Center. In his presentation, Mesarovic spoke of recent projects that had been initiated to apply this 'multilevel approach' to emerging problems in urban systems analysis and resource management. A member of the seminar audience was Eduard Pestel, a member of the Executive Committee of the Club of Rome as well as a highly influential member of the European intellectual-scientific community.

The Mesarovic seminar came at a time when Pestel and other leading members of the Club of Rome were beginning to explore the possibility of a follow-up to the MIT project. They would soon conclude that the Forrester/Meadows group had achieved most of their objectives, and that reactions to *The Limits to Growth* had pointed the way towards new initiatives. A follow-up project was needed that would:

1. attempt to represent the world as a system of interdependent regions,
2. focus on the development of recommendations that would be of more direct policy relevance, and
3. attempt to incorporate 'hard data' as well as the theories and frameworks of relevant disciplines more explicitly.

One possibility, of course, would have been to undertake the follow-up at MIT, building on the experience of the Forrester/Meadows group. But several considerations dictated otherwise. Systems dynamics was viewed with skepticism— or worse— in many traditional scientific circles. In some quarters, the MIT group had come to be viewed as zealots who advocated 'no growth'. Thus, a follow-up project would gain wider acceptance if it were not freighted with the burden of debate and controversy, sometimes personalized and acrimonious, which had come to be associated with Forrester, Meadows, and their colleagues. Following a series of discussions, Mesarovic and Pestel decided to undertake a project that would offer a more disaggregated view of the world and profit from the experiences of the earlier effort. Mesarovic's multilevel approach would be used as the basis for an entirely new model.

The Mesarovic/Pestel project is particularly interesting from an organizational standpoint. Instead of a single group, two teams, located at Case Western Reserve and Hannover's Technical University, worked on different parts of the model in parallel. The demographic, environmental, and portions of the economic sectors of the model were developed in Hannover (ultimately the environmental sector was not included in the model). The Cleveland group was responsible for energy, agriculture, international trade, and, ultimately, for integrating the several parts of the model into a coherent whole. Integration was complicated, not only by the distance between the two groups but also by the very different structures of the respective sectors. Instead of using a common paradigm, such as system dynamics, the team tried to make each sector resemble, insofar as was possible, state-of-the-art work in the relevant academic disciplines. Ultimately, the core group systems programmer, Thomas Shook, developed a specialized software package, called MODEL BUILDER, which facilitated submodel integration and on-line interaction with the model. Although somewhat complex, this software is still regarded as the state of the art and an outstanding contribution to the modelling field.

The Mesarovic/Pestel model was first unveiled at a meeting for high-level policy makers sponsored by the Woodrow Wilson International Center for Scholars in Washington, D.C. (A public presentation of the World3 model had occurred in this same forum.) Subsequently, the model was the focus of discussion at IIASA's first global modelling conference. After *The Limits to Growth,* Mesarovic and Pestel decided that presentation of results to a scientific audience should precede publication of a popularized version. After the IIASA meeting, Mesarovic, Pestel, and other team members delivered papers on the project at scientific meetings throughout the world. Because of careful attention to the concerns of the scientific community, both in the model structure and in the mode of presentation, the results of the project were received rather favourably. The popular work describing the project, *Mankind at the Turning Point* (Mesarovic and Pestel, 1974), was released in Washington and, shortly thereafter, at the Berlin meeting of the Club of Rome in the fall of 1974.

Mankind at the Turning Point presents a complex model, a vision of future crises, and a set of recommended solutions. It deliberately avoids simple conclusions or recommendations. The style of the book is somewhat more academic than that of *Limits*. For these reasons, (and, no doubt, some which have not occurred to us) *Mankind* has not received the degree of popular attention that was accorded to *Limits,* especially in the United States.

The purpose of the model

Here is the way the authors describe their purpose in the preface to *Mankind at the Turning Point* (Mesarovic and Pestel, 1974, *p.ix*):

> The scientific foundation for the research ... was laid nearly three years ago when the authors decided to develop a project concerned with the analysis of global issues which would realistically take into account the diversity of the many different world regions and would deal with issues concretely, rather than in abstract terms. We hoped thus to furnish political and economic decision makers in various parts of the world with a comprehensive global planning tool, which could help them to act in anticipation of the crises at our doorstep and of those that loom increasingly large in the distance, instead of reacting in the spirit of short-term pragmatism.

In the next chapter of this book Mesarovic once again emphasizes the very broad scope of his model and the breadth of concerns which motivated its creation:

> The single most important problem which a global model should try to analyze is the so-called *world problematique* identified by the Club of Rome; namely, that mankind *is* entering a new era in its history in which a global society is emerging. Obviously, this society is emerging not in an organized manner or by 'design' but as an inevitable consequence of a finite globe and the interdependence of the parts within the global system. The basic question is: *How should human affairs be managed so that the progress—even if at a slower pace—can continue* during the period of transition?

Two important themes stand out clearly in these statements of purpose (and contrast sharply with the purposes of World3):

1. Insofar as is possible, the model was intended to address the problematique in all its complexity and to permit exploring a host of sectoral and regional problems from a global perspective. (Later versions of the model have been used extensively for this purpose.)

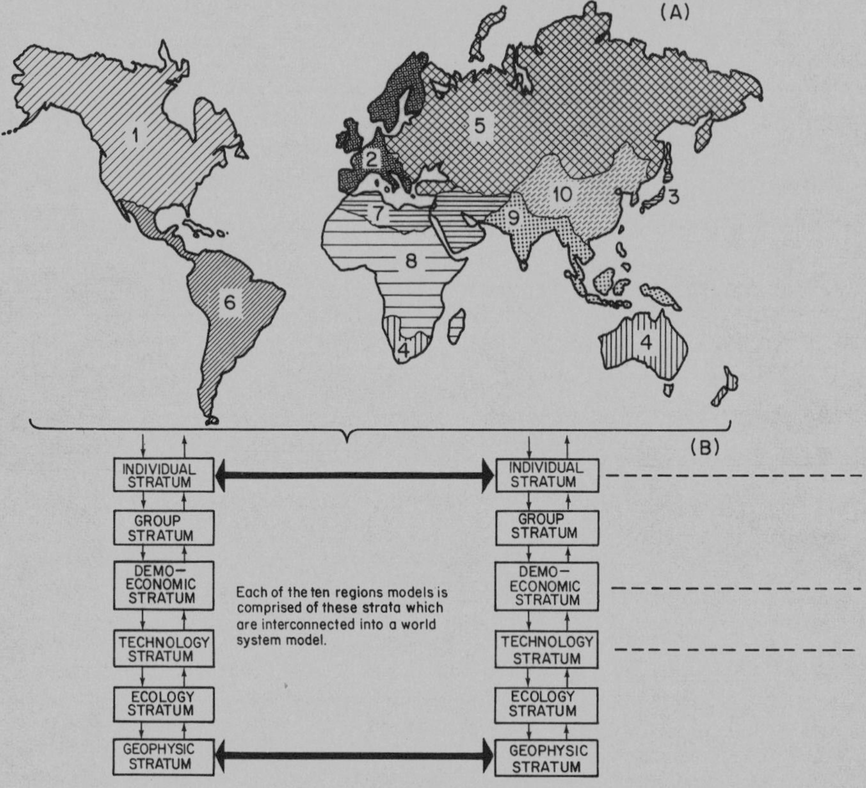

Figure 5. The basic structure of the Mesarovic/Pestel model. (From *Mankind at the Turning Point* by Mihajlo Mesarovic and Eduard Pestel. Copyright © 1974 by Mihajlo Mesarovic and Eduard Pestel. Reprinted by permission of the publisher, E.P. Dutton)

2. The model was intended for actual use by decision makers as a practical planning and policy-assessment tool.

These themes have continued to motivate the activities of the Mesarovic/Pestel project throughout its history. 'How can we convince decision makers to use the model?' and 'How can they use it most effectively?' have been overriding concerns.

Structure of the model

Because the Mesarovic/Pestel model uses different paradigms for different sectors and, to some degree, has a structure that can be adapted flexibly to different problem areas, it defies easy description. Probably the best general overview is provided by Figure 5 (from *Mankind at the Turning Point*), which is divided into four parts:

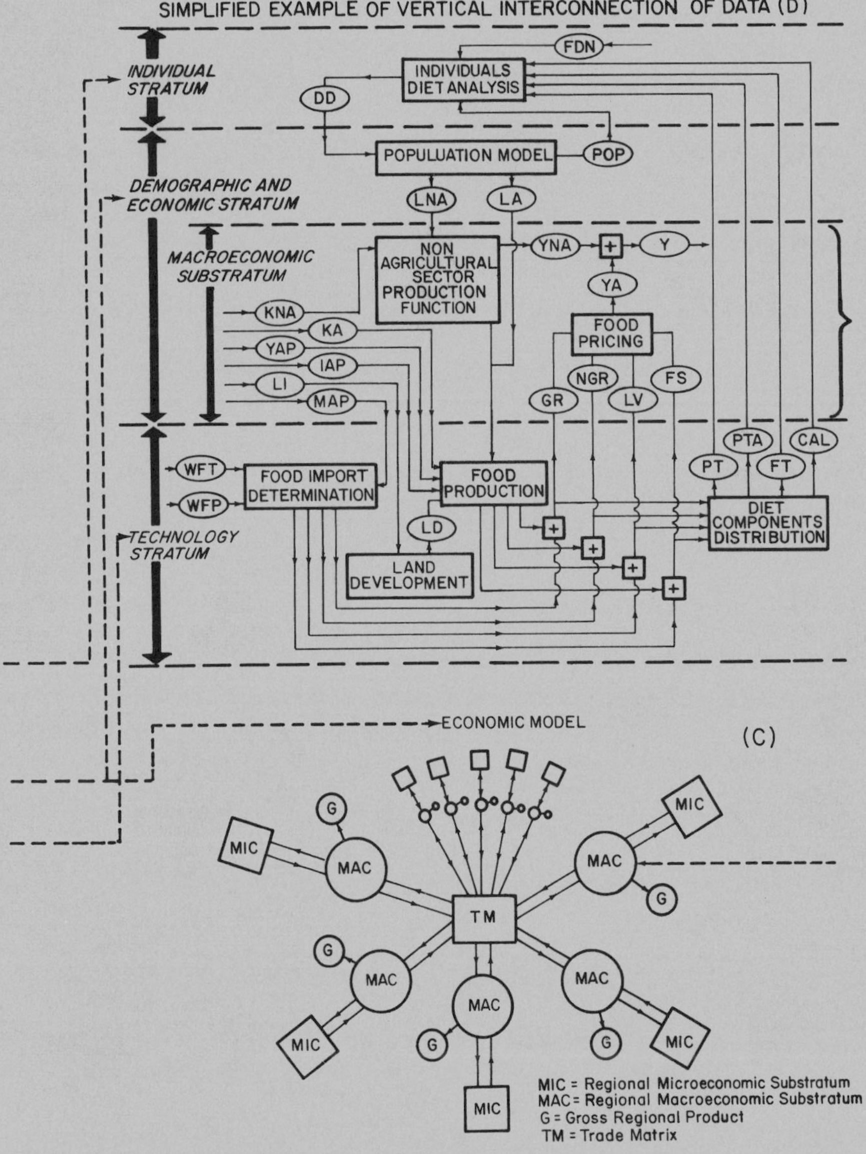

Figure 5 (C and D). The basic structure of the Mesarovic/Pestel model (from Mesarovic and Pestel, 1974, p. 45)

(A) regionalization of the model,
(B) the 'multilevel' structure of the regional submodels,
(C) a more detailed description of the structure, applied, in this case, to problems of agriculture and food production,
(D) a depiction of the linkages among the regional economic strata.

Each of these parts will be described briefly.

The basis for *regionalization* is evident from the map in Figure 5. The scheme takes into account both geographic proximity and a degree of similarity in culture, outlook, politics, and economic development. In later versions, several regions (including Europe, South Asia, and the Middle East) have been further divided, but the basic idea has remained intact. The *multilevel approach* (Figure 5, part B) originated in Mesarovic's earlier theoretical writings. This concept has served as an enormously useful heuristic for organizing complex models in a variety of areas. It will not necessarily be the case that there is a specific, well-defined submodel uniquely associated with each stratum illustrated here. Rather, the scheme provides a basis for broad conceptualization, grouping of submodels, and relating them to an overall structure. This overall conceptualization of the model has remained unchanged throughout its numerous stages of evolution.

A second important feature of the approach—division into causal and decision strata—is illustrated in Figure 6. This figure is often used by the modellers to illustrate the way in which the model is intended to be used. *Causal-stratum* phenomena are relatively deterministic, subject to scientific investigation, and governed by natural law. *Decision-stratum* phenomena involve norms, options, and choices and probably should not be modelled at all. Instead, the best approach is for a real decision maker to insert himself or herself into the process and interact directly with the computer. Consequently, the Mesarovic/Pestel model is usually run as a man-machine system, with human actors entering policy decisions or future scenarios and the computer calculating the results throughout the system.

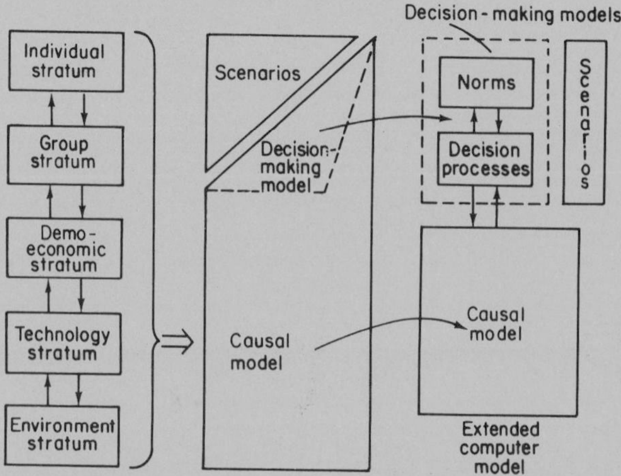

Figure 6. The multilevel approach to model building. (From *Mankind at the Turning Point* by Mihajlo Mesarovic and Eduard Pestel. Copyright © 1974 by Mihajlo Mesarovic and Eduard Pestel. Reprinted by permission of the publisher, E.P. Dutton)

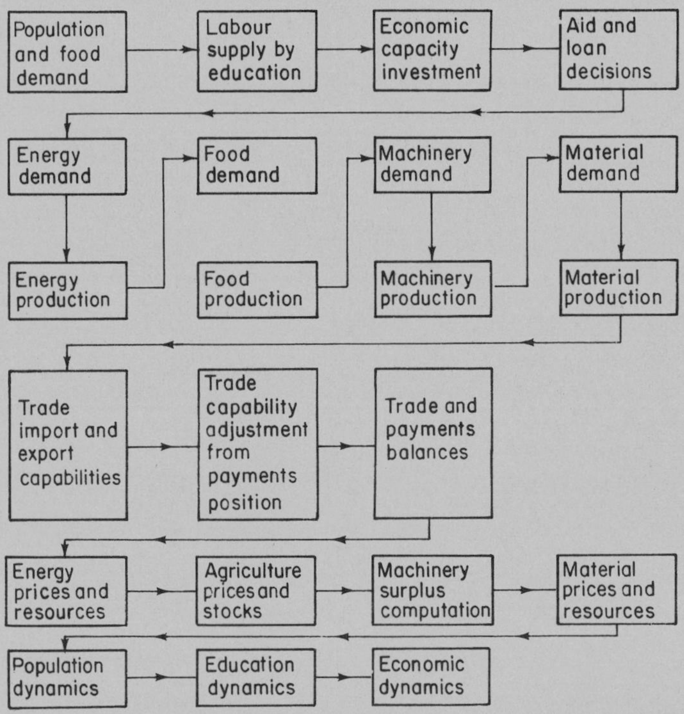

Figure 7. The sequence of computations in the world integrated (Mesarovic/Pestel) model. (Reprinted by permission of the publisher from *World Modelling* by Barry B. Hughes. Lexington, Mass.: Lexington Books, D.C. Heath and Company, Copyright ©1980, D.C. Heath and Company)

The more detailed description in Figure 5, part D, focuses on the components of the model used to address problems of agricultural production and nutrition. The boxes represent particular components within which calculations are made. The lozenge-shaped symbols with capital letters stand for specific variables. (For example YA stands for the output of the agricultural sector measured in dollars.) This type of diagram is quite commonly used by engineers with an electrical- or systems-engineering background. Although it is not immediately evident, this diagram also illustrates one other unique feature of the model. Flows of commodities are accounted for in both monetary and physical terms. Thus, the model attempts a synthesis of ecological/system dynamics perspectives (emphasizing the flows of 'real' things) with those of traditional economics (emphasizing flows of money).

The section of the diagram labelled 'the economic model' illustrates the way in which microeconomic and macroeconomic perspectives are combined. Regional interaction occurs only at the highly aggregated macroeconomic level. But much more detailed within-region computations may also be provided by the model. (This can of course be helpful to a decision maker focusing on the concerns of a particular region). This detail is customarily not taken into account when the model is used to view problems from a global perspective.

A somewhat simpler picture of the model, illustrating the sequential flow of computation, has been provided by Barry B. Hughes in a recent work (Hughes, 1980) (see Figure 7). This sequence is, of course, repeated for each region. The computations relating to regional interaction are those involving trade capabilities and balance of payments.

Although the Mesarovic/Pestel model, as it has been depicted here, looks very different from World3 it is not greatly different in a mathematical sense. Most of the model could be portrayed using the system dynamics notation. However, seemingly superficial differences in the way a model is presented can often turn out to be very important. In this case, the method of presentation was specifically designed to appeal to relatively conservative scientific audiences. In general, this strategy was quite successful in defusing the criticism that had been directed at World3.

Results

Owing to the emphasis on using the model as a planning and decision-making tool, Mesarovic and Pestel have been reluctant to emphasize results. 'Results' are for the user to determine, they argue, based on the incorporation of his own assumptions into the model. Nonetheless, it is useful to present two sets of 'results' from *Mankind at the Turning Point* that illustrate particularly important and recurrent themes of the project:

1. Cooperation is better than confrontation, even when one views the world from a narrowly defined and self-interested perspective.
2. Delays in addressing critical global issues can be disastrous, even deadly.

Figure 8 compares the GNP for several world regions under contrasting assumptions of a very low fixed level of oil price and an 'optimal' level. It turns out that the 'optimal' price is not only better for the oil producers but also for the consumers (at least those in the developed world). Price increases permit a gradual adaptation to scarce oil in the developed regions rather than a catastrophic economic decline when the resource is suddenly exhausted. Further analysis, not pictured here, shows that prices above the 'optimal' level also leave everyone worse off.

Figure 9 depicts the effects of a five-year delay in effecting a population-control policy aimed at producing population equilibrium. Curves 1 and 3 show the effects of the policy implemented in 1990. Curves 2 and 4 present similar data for 1995. The five-year delay has negligible consequences for the total population, but the difference in mortality is significant. The number of additional child deaths resulting from the five-year delay could amount to as many as 150 million.

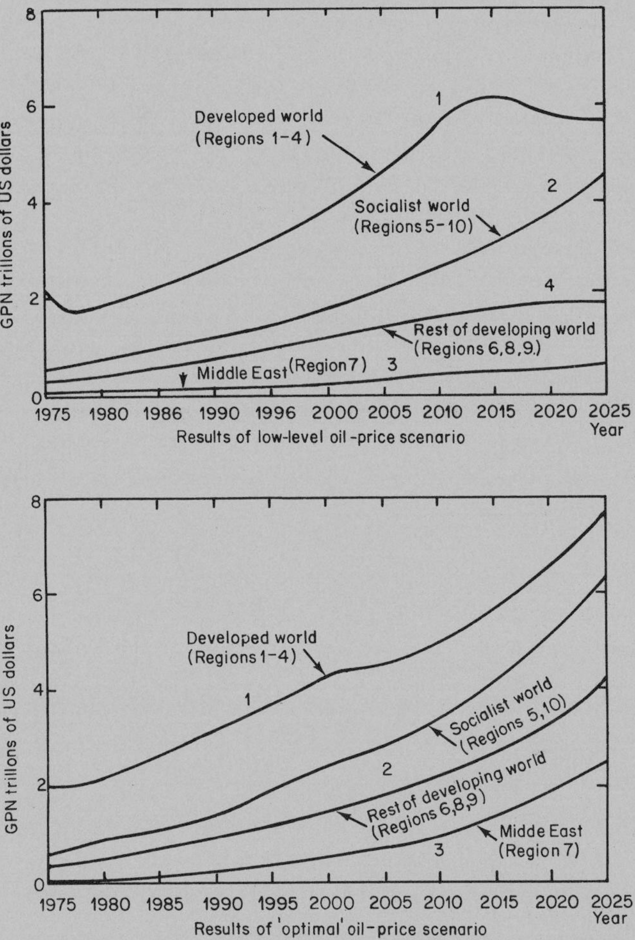

Figure 8. Comparison of long-term world development for a fixed low-level oil-price scenario and an optimal oil-price scenario. (From *Mankind at the Turning Point* by Mihajlo Mesarovic and Eduard Pestel. Copyright © 1974 by Mihajlo Mesarovic and Eduard-Pestel. Reprinted by permission of the publisher, E.P. Dutton)

The general conclusions presented in *Mankind at the Turning Point* go well beyond the specific analyses presented here. In a very real sense, they are based, not on the model, but on the insights the authors gained from using it. The latter type of conclusions are often the most interesting and may be the most significant to emerge from global modelling projects.

Among the conclusions presented (most of which derive from the authors' mental models and not directly from the computer model) are these (as paraphrased from Mesarovic and Pestel, 1974, pp. 143-147, emphases as in the original):

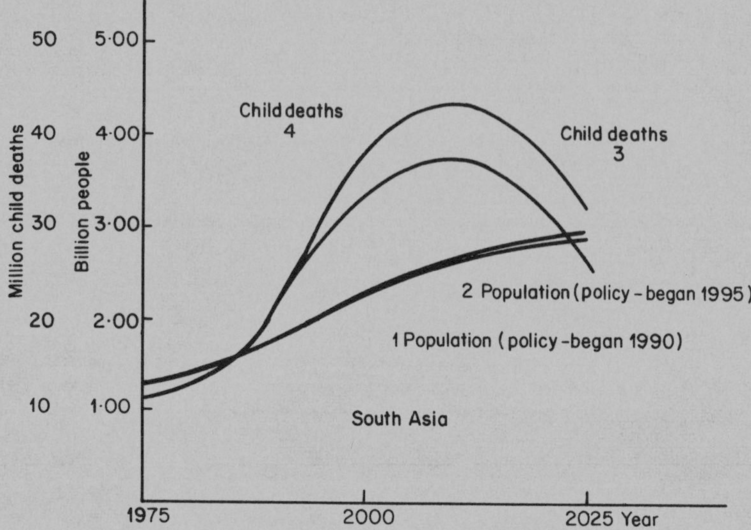

Figure 9. 'Deadly consequences of population delay in South Asia.' (From *Mankind at the Turning Point* by Mihajlo Mesarovic and Eduard Pestel. Copyright © 1974 by Mihajlo Mesarovic and Eduard Pestel. Reprinted by permission of the publisher, E.P. Dutton)

A. *About the current* (circa 1974) *global crises:*

1. *The current crises are not temporary,* but rather reflect a persistent trend inherent in the historical pattern of development.
2. *The solution of these crises can be developed only in a global context and a long-term basis* with full and explicit recognition of the emerging world system.
3. *The solutions cannot be achieved by traditional means* confined to an isolated aspect of the world system, such as economics. What is really needed is nothing short of a complete integration of all strata in our hierarchical view of world development.
4. *It is possible to resolve these crises through cooperation rather than confrontation;* indeed, in most instances, cooperation is equally beneficial to all participants. The greatest obstacles to cooperation are the short-term gains that may be obtained through confrontation.

B. *Changes needed on the societal level:*

1. People must realize that *counterproductivity* is the ultimate consequence of any action confined solely to short-term considerations.
2. The *futility of narrow nationalism* must be appreciated and taken as an axiom in the decision-making framework. *Global issues can be solved only by concerted global action.*

3. A practical international framework must be developed in which the *cooperation* essential for the emergence of a new organic growth path will become a matter of necessity, rather than being left to goodwill and preference.
4. People must realize the *overriding importance of the long-term global development crises,* and be willing to place this highest on the agenda of issues to be dealt with explicitly by national governments and international organizations.

C. *Lessons for the emergence of a new global ethic:*
1. *A world consciousness* must be developed through which every individual realizes his role as a member of the world community. Famine in tropical Africa should be considered as relevant and as disturbing to a citizen of Germany as famine in Bavaria.
2. *A new ethic in the use of material resources* must be realized that will result in a style of life compatible with the oncoming age of scarcity. One should be proud of saving and conserving, rather than of spending and discarding.
3. *An attitude towards nature must be developed based on harmony rather than conquest.*
4. If the human species is to survive, man must develop a *sense of identification with future generations* and be ready to trade benefits to the next generations for benefits to himself. If each generation aims at maximum good for itself, *homo sapiens* is as good as doomed.

Further developments

The publication of *Mankind at the Turning Point* marked the beginning of an intensive period of activity by the group during which the model was both refined and disseminated. Most of this work was carried out by the group in Cleveland under Mesarovic's tireless leadership. Although the composition of the group changed gradually, Thomas Shook continues to be involved and Barry Hughes played a major role until 1979. John Richardson left the group in 1975, but continues to be active as an observer and critic of the field. Peter Clapham now works as an independent consultant in Cleveland.

Mesarovic's group has now developed at least six distinct refinements of the world integrated model. In addition, they have developed the capability of presenting interactive computer demonstrations throughout the world using satellite phone patches to their computer in Cleveland. A recent publication of the project described twenty-four such presentations, many to prime ministers and other high-level officials in eighteen countries of the world. While continuing an extensive dissemination program, Professor Mesarovic is presently in the process of forming a new group and further refining the structure of the model. Professor Pestel is serving as a provincial Minister for Lower Saxony while continuing to be involved in the Club of Rome and international scientific affairs.

The principal references

Main publication

Mesarovic, M.D., and Pestel, E. (1974). *Mankind at the Turning Point,* Dutton, New York.

Other publications describing the model and its application

Hughes, Barry B. (1980). *World Modelling,* D.C. Heath, Lexington, Mass.

Mesarovic, M.D., *et al.* (1970) *Theory of Hierarchical Multilevel Systems,* Academic Press, New York.

Mesarovic, M.D., and Pestel, E. (1972). 'A goal seeking and regionalized model for analysis of critical world relationships: the conceptual foundation', *Kybernetes,* **1**, 79-85.

Mesarovic, M.D., and Richardson, J.M., Jr. (1973). 'A proposed strategy and recommendations for dealing with the world food crisis', in *Growth and Its Implications for the Future* (4 vols.), Committee on Merchant Marine and Fisheries, 93rd Congress, US Government Printing Office, Washington, D.C.

Mesarovic, M.D., and Pestel, E. (Eds.) (1974). 'Multilevel computer model of world development system', *Proceedings of the Symposium held at IIASA,* Vols. I-VI, SP 74-1/6.

Richardson, J.M., Jr. (1975). 'Hierarchical Models of Very Large Problems: Dilemmas, Prospects and an Agenda for the Future', in *Large Scale Dynamic Systems* (Ed. Clyde Smith), NASA.

Mesarovic, M.D., and Hughes, B.B. (1978). 'Testing the Hudson Institute scenarios: is their optimism justified', *Futurist,* **XX**, I:6, November/December 1978.

Mesarovic, M.D. (1979). 'World modeling and its potential impact', in *Growth in a Finite World* (Ed. Joseph Grunfield) Franklin Institute Press, Philadelphia, Pa.

Reviews of Mankind at the Turning Point

Boulding, K., 'Conditional optimism about the world situation, *Science,* **187**, 1188-9.

Economist, 'Another crack at doom', July 1975.

Naughton, J. (1975). 'A bold, bad computer', *Listener,* 27 March, 1975.

The Bariloche Model

Implicit in every global modelling project is the idea that some images of the future are preferable to others. It is impossible to build a global model without reflecting on questions like

> What is meant by 'quality of life'?
> What future global order would be best for humankind?
> Is this global order attainable?
> What are the principal obstacles that impede its attainment?
> What policies and institutional arrangements would most facilitate attaining a new global order in which the quality of life is optimized?

Among the major projects, however, only the Bariloche group has taken these questions as the principal focus of their model-building activities. The result of their efforts is a model that uses the most powerful, abstract mathematics (the mathematics of optimization) and that is, at the same time, the most explicitly ideological of the seven models we are considering.

Like the work of Mesarovic and Pestel, the impetus for the Bariloche model came from discussions of World3. The idea of the model emerged from a 1970 meeting in Brazil, sponsored jointly by the Club of Rome and the Instituto Universitario Pesquisas de Rio de Janeiro. Preliminary results emerging from the work of the MIT team were presented and discussed at this meeting. The reaction of the largely Latin American participants was mostly negative and presaged criticisms that have since been made by developing-world leaders in many forums. The criticisms focused on three major concerns:

1. Predictions based on an extrapolation of present structural arrangements in the global system are uninteresting; what should be examined is the potential for fundamental changes in values and institutions.
2. The view that global crises will occur in the future reflects a parochial, developed-world perspective. For two-thirds of the world's population, crises of scarce resources, inadequate housing, deplorable conditions of health, and starvation are already at hand.
3. Policies oriented towards the attainment of a 'state of global equilibrium' in the near future will ensure that the present inequities and disparities in the global system are perpetuated.

As a result of the Rio de Janeiro meeting, a group of Latin Americans committed themselves to developing a 'normative model' that would be concerned 'not with predicting what will occur if the contemporary tendencies of mankind continue, but rather with sketching a way of arriving at the final goals of a world liberated from backwardness and misery' (Bariloche Group, 1976, p. iii). A committee composed of Carlos A. Mallman, Jorge Sábato, Enrique Oteiza, Amilcar O Herrera, Helio Jaguaribe, and Osvaldo Sunkel was established to outline the general aims of the project and to effect its implementation. The project was carried out at Argentina's Fundacion Bariloche (hence the most commonly used name for the model), with principal support provided by the International Development Research Centre in Ottawa, Canada.

The Bariloche model was first presented at IIASA in the fall of 1974. However, it was not until 1976 that a popular description *Catastrophe or New Society?* was available in English. This description, published as a report by the International Development Research Centre in Ottawa, Canada, presented the structure of the model and the principal conclusions of the project with force and clarity. But, unfortunately, it has not received the wide dissemination of *The Limits to Growth* and *Mankind at the Turning Point*. Subsequently, reports on the model were published in German, French, Japanese, and Dutch, and one is forthcoming in Portuguese.

The Bariloche model has probably achieved its greatest impact through the United Nations. A fifteen-region version of the model was developed under the auspices of the International Labor Organization (ILO). Results based on this version of the model were presented in a widely read ILO publication (Hopkins and Scolnik, 1976) and provided a major contribution to the World Employment Conference, held in Geneva in 1976. Among third-world scientists and decision makers, as well as those who are concerned with the 'basic needs' approach to development, the Bariloche model is by far the most popular of the major global models.

The purposes of the model

The appropriate place to begin discussing the Bariloche model is with the authors' vision of an ideal society. For it was with the goal of helping to bring about this society on earth that the Bariloche team pursued its task.

The words 'egalitarian, fully participatory and nonconsuming' (Herrera, Scolnik, *et al.*, 1976) best characterize this society. The team envisioned a world 'fully liberated from underdevelopment and misery', in which human needs and human rights, rather than the desires to consume and accumulate wealth, would become the basis for resource allocation. 'But it is not sufficient simply to describe an ideal society' the authors emphasized, 'it is necessary also to demonstrate its material viability.'

> Thus, we must start by showing beyond all reasonable doubt that, for the foreseeable future, the environment and its natural resources will not impose barriers of absolute physical limits on the attainment of such a society. Secondly, it must be demonstrated that different countries and regions of the world (particularly the poorest) could reach the goals we advocate in a reasonable period of time, starting from the current situation as regards the availability of capital, manpower, land, demographic trends, etc. (Herrera, Scolnik, *et al.*, 1976, p.8)

The more specific goals that guided the project and the development of the model followed clearly from this objective:

> To attain the first objective—to demonstrate that absolute physical limits do not exist for the foreseeable future—an analysis was undertaken of the current situation in nonrenewable resources, energy and pollution. A mathematical model was built to handle the second objective: establishing that all countries or regions of the world could move from their present situations to the postulated goals in a reasonable time. Thus, the conceptual model is a proposal for a new society and the mathematical model is the instrument through which its material viability is explored. (Herrera, Scolnik, *et al.*, 1976, p.8. Reproduced by permission of the International Development Research Centre, Ottawa, Canada)

The analysis of the resource situation was basically a static compilation of current resource data and a theoretical argument against resource scarcity.

> Except for a few cases, the vast volume of the earth's is mineral resources, although used, continue to be part of the earth's resources as if they had never been extracted. They can be disseminated throughout the earth or spread over the oceans; they can be chemically combined with other elements; they cannot be destroyed. (Amilcar Herrera in Bariloche group, 1976, p.66)

On the basis of this analysis, the authors concluded that resources and environmental limits are not a serious problem. Therefore these factors were not included in the computer model.

Thus, the purpose of the Bariloche model was to provide explicit, concrete answers to the first three fundamental questions posed above:

1. What is meant by 'quality of life'?
2. What future global order would be best for humankind?
3. Is that global order attainable?

In their published writings, but not specifically in their model, the team addressed the final two questions as well:

4. What are the principal obstacles to attaining a new global order?
5. What policies and institutional arrangements would facilitate attaining this new order?

The structure of the model

Figure 10 shows three major components of the Bariloche model, which can best be described as:

(A) *The productive economic sectors*
- (1) food production (nutrition)
- (2) housing
- (3) education
- (4) other consumption goods and services
- (5) capital goods

(B) *Factors affecting the level of economic productivity in each year*
- (1) labour
- (2) capital

(C) *Population,* which includes not only the size of the population but also considerations of
- (1) life expectancy at birth
- (2) health

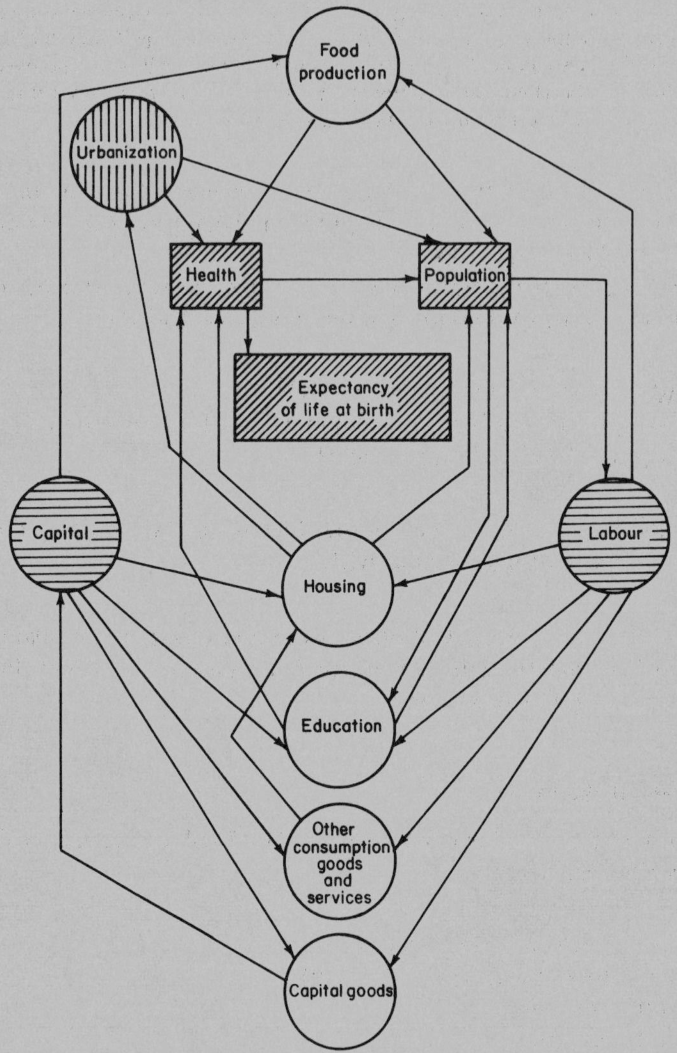

Figure 10. The basic structure of the Bariloche model. (from Graciela Chichilnisky, 'Latin American world model: theoretical structure and economic sector', in Bariloche Group, 1976, p. 126)

The model operates by allocating capital and labour to the various productive sectors (food, housing, education, etc.) in whatever way is necessary to maximize life expectancy at birth.

In the judgement of the modellers,

> life expectancy at birth is, without doubt, the indicator that best reflects the general conditions of life, regardless of country. Its value is a function of the extent to which the basic needs are satisfied and of other factors, such as

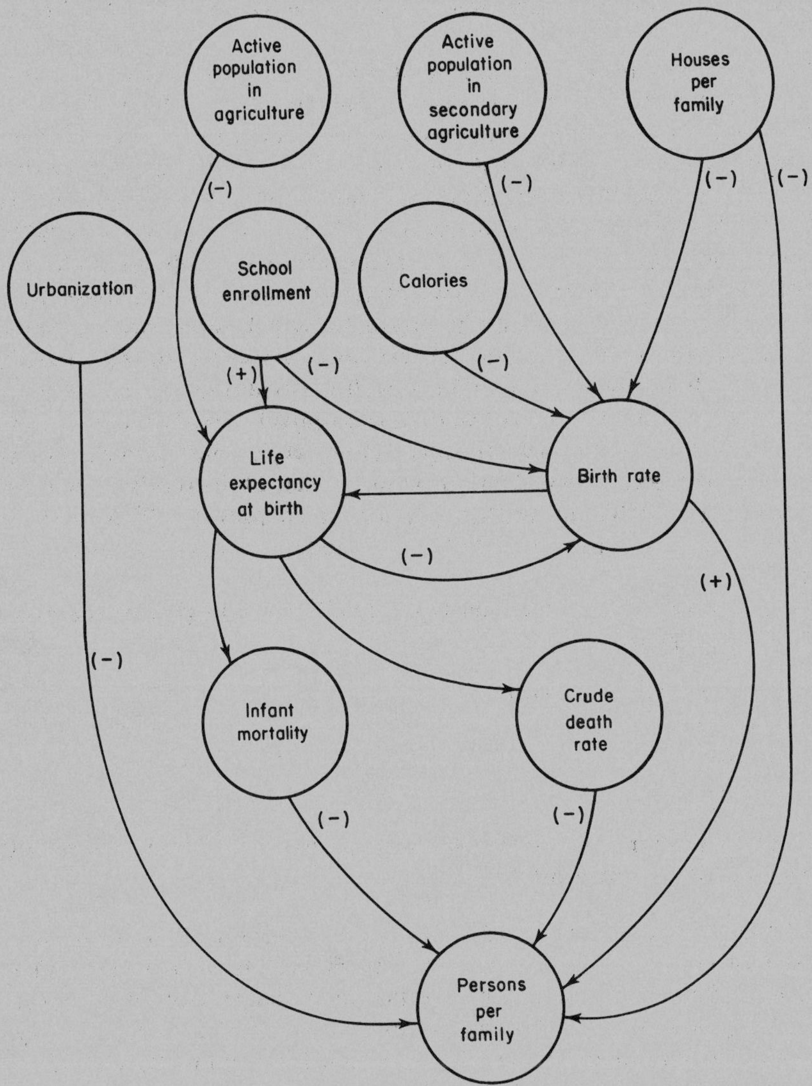

Figure 11. The principal variables and interrelations in the population submodel of the Bariloche model. (Reproduced by permission of the International Development Research Centre, Ottawa, Canada, from Herrera, Scolnik *et al.*, 1976, p. 50)

urbanization, that most affect the lives of members of the community. (Herrera, Scolnik, *et al.*, p. 53, 1976)

A somewhat complicated mathematical optimization procedure is used to ensure that the available resources will be used most productively to achieve this goal and that no resource is overused. The result of this procedure is obviously not a picture of how the world *will* be but how it *could* be if all allocation decisions actually were directed optimally towards the life-expectancy objective.

As can be seen from the diagram, the output of the productive sectors is wholly determined by available amounts of capital and labour. A classical Cobb-Douglas production function is used to calculate this relation. The size of the population for each year, as well as the overall state of health, is determined, essentially, by the degree of satisfaction of basic human needs.

The population submodel is designed to 'identify the factors in economic and social development that influence the demographic evolution of society'. Figure 11 is a diagram of the principal variables and interrelations in this submodel. As can be seen from the diagram, the model incorporates several variables that express the relations between life expectancy at birth, population growth, and basic human needs. The modellers circumspectly emphasize that the relations are based on statistical (i.e., descriptive) analysis and 'should not be confused with causal relationships'. However, in their judgement, 'the results are very satisfactory, from the point of view of the precision achieved, and make it possible to predict population changes resulting from the socioeconomic variables considered in the model' (Herrera, Scolnik, *et al.*, 1976, p. 51).

The structure of regionalization is the final important part of the overall model. Four regions are identified:

1. the developed countries,
2. Latin America,
3. Africa,
4. Asia.*

This regionalization reflects the primary concern of the modellers with the problems of the Third World.

> International trade is included in the model in the form of the aggregate value of imports and exports of regions, disaggregated by sectors. The sectors affected are capital goods, other goods and services, and goods. Each region is almost self-sufficient in education and housing, the very small external inputs required being included in capital and consumer goods. (Herrera, Scolnik, *et al.*, 1976, p. 43).

Results

The model runs for a hundred-year period from 1960 until 2060. A continuation of present policies is assumed for the first twenty years. The modellers assume that after 1980 the policies designed to give the highest priority to basic human needs are adopted. The optimal mix of policies is calculated by the model. For illustrative purposes, we present both graphic and

*A later version of the model, developed by Hopkins and Scolnik (1976), used the same regional breakdown as the UN world model (see below).

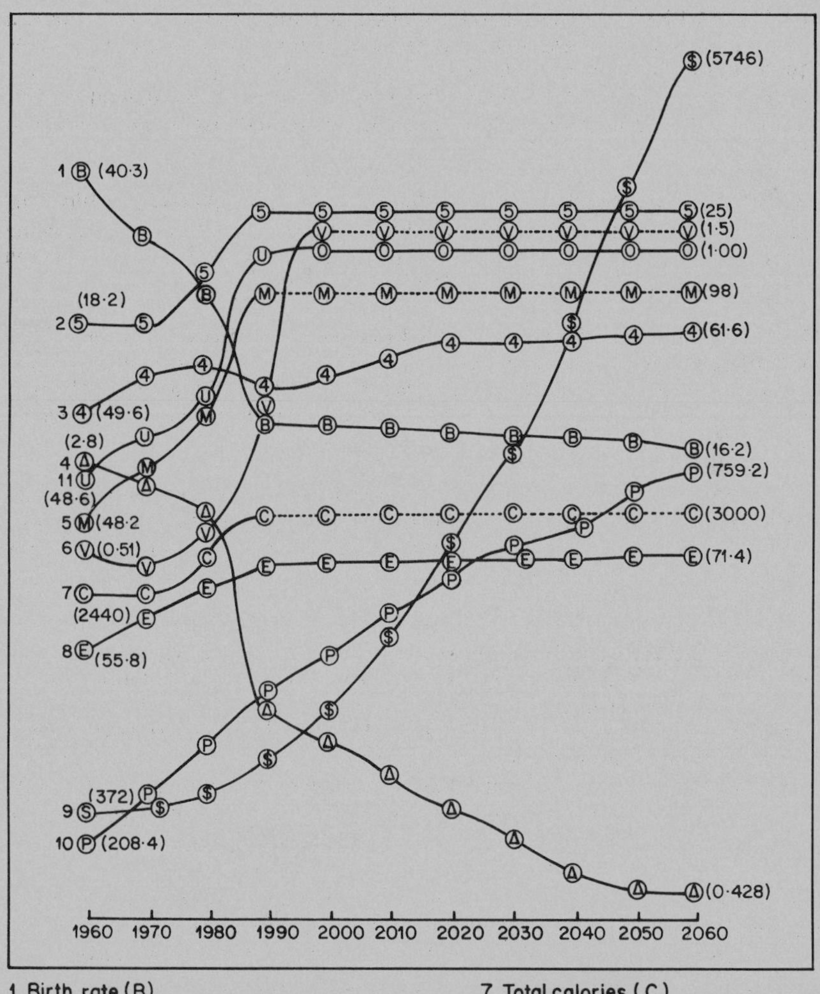

1 Birth rate (B)
2 Percentage of GNP allocated to sector 5 (5)
3 Percentage of GNP allocated to sector 4 (4)
4 Population growth rate (Δ)
5 Enrollment (M)
6 Houses per family (V)
7 Total calories (C)
8 Life expectancy (E)
9 GNP per capita in 1960 dollars ($)
10 Total population (P)
11 Urbanization (U)

Figure 12. The time periods and conditions required for Latin America to satisfy basic needs to given levels. (Reproduced by permission of the International Research Centre, Ottawa, Canada from Herrera, Scolnik, *et al.*, 1976, p. 88)

tabular results for Latin America in Figure 12 and Table 2. The graph and table show the results of model runs for the main demographic and economic indicators. Similar results are reported for the other regions.

The modellers' summaries of these results are brief, informative, and straightforward (Herrera, Scolnik, *et al.*, 1976):

LATIN AMERICA: The general evolution of Latin America, if the proposed policies are applied, would make it possible to fulfill the basic needs in the early 1990s. Regarding food, it would be necessary to develop relatively little extra land for agricultural purposes .... Before the turn of the century, it would already be possible to have a stock of food sufficient for 1 year, and that level is maintained until 2060.

It is, perhaps, the population figures that show the most interesting results. The growth rate, which was 2.8% in 1960, decreases as general well-being improves .... Population stabilization is therefore approached. Total population rises from 208.4 million in 1960 to 759 million in 2060.

In conclusion, Latin America could adequately satisfy the basic needs of the whole population within one generation from the implementation of the proposed policies. Subsequently, Latin America could improve its general level of well-being as indicated by increasing the proportion of resources allocated to consumption (pp. 87-88).

AFRICA: 'Africa also satisfies its basic needs, but in a longer time period than Latin America, the target being reached in 2008, (p. 89). It should be noted, however, that 'the quality of the housing stock will still be substantially below that of the developed countries' (p. 91).

ASIA: In Asia the problems are more serious and basic needs are not satisfied to the desired levels. Shortfalls exist in both food and housing, with a serious food shortage emerging after the turn of the century. A more optimistic sign is that basic educational needs can be satisfied in the year 2040. 'The failure to attain the satisfaction of basic needs to the desired levels is reflected in the demographic indicators. The rate of population growth is reduced slowly and the population increases fivefold in 80 years, reaching 7840 million in 2040. Life expectancy at birth improves but is always below the levels in other regions. Also,

Table 2. Evolution of the main economic and demographic and health indicators for Latin America. (Reproduced by permission of the International Development Research Centre, Ottawa, Canada. From Herrera, Scolnik, *et al.*, 1976, p. 89)

	1960	1980	2000	2020	2040	2060
Economic indicators						
GNP per capita	372	530	1107	2247	3822	5746
Investment rate (% GNP)	18.2	21.2	25	25	25	25
Consumption (% GNP)	49.2	55.8	54.8	59.8	60.6	61.6
% GNP allocated to food	21.2	14.21	10.63	7.69	6.3	5.34
Demographic and health indicators						
Population growth rate (%)	2.8	2.6	1.27	0.89	0.56	0.43
Total population (millions)	208.4	350.6	486.3	601.2	693.9	759.2
Life expectancy (years)	55.8	65.8	70.24	20.75	71.04	71.38
Crude mortality rate	14.7	7.02	5.91	8.53	11.56	12.03
Birth rate	40.36	30.04	18.34	17.57	17.07	16.22
Infant mortality	115	37	24	22.5	21.7	20.88
Persons per family	4.5	4.9	3.5	3.3	3.1	3

infant mortality does not compare favorably with Latin America and Africa' (p. 93).

DEVELOPED COUNTRIES: In the developed countries, the authors conclude that high levels of well-being can be reached even if their material growth rate is drastically reduced in the future. Work can be reduced and leisure time increased, while a growth rate is maintained that will preserve and continuously improve the physical and human environment.

Further developments

As noted above, the Bariloche model, expanded to fifteen regions, contributed significantly to the background document for the 1976 World Employment Conference. The emphasis on the basic needs of health, housing, nutrition, and education were in accord with the ideals of the ILO. With the assistance of the model and one member of the Bariloche team, conference planners were able to formulate a clearer and more precise statement of those ideals than might otherwise have been possible. At about the same time, it became necessary for political reasons to conclude activities in Argentina and move the model to a new home in Brazil. For more than two years the ASDELA group, under Hugo Scolnik's direction, continued the work of retrieving the model and publishing its message. This effort, which was located at Candido Mendez University, concluded in 1979.

By the summer of 1980, there was no team specifically responsible for maintaining and developing the Bariloche model further. Some team members continued to be active at international scientific meetings and in UN and Club of Rome activities. But the humane and just global order that the model describes so well remains a goal of concerned individuals, nations, and international organizations throughout the world.

The principal references

Bariloche Group (1976). 'The Latin American World Model', *Proceedings of the Second IIASA Conference on Global Modeling* (Ed. Gerhart Bruckmann), CP-76-8, International Institute for Applied Systems Analysis, Laxenburg, Austria.

Herrera, A.D., Scolnik, H.D., *et al.* (1976). *Catastrophe or New Society?: A Latin American World Model,* International Development Research Centre, Ottawa, Canada (short version).

German: *Grenzen des Elends,* S. Fischer Verlag, Frankfurt, 1977 (full version).

French: *Un monde pour Tous,* Presses Universitaires de France, Paris, 1977 (full version).

Japanese: Diamond Inc., Tokyo, 1976 (short version).

Dutch: *Het Bariloche- rapport voor de Club van Rome,* Het Spectrum, Utrecht, 1978.

Hopkins, M.J.D., and Scolnik, H.D. (1976). 'Basic needs, growth and redistribution: a quantitative approach', in *Basic needs and national employment strategies,* World Employment Conference, Background Papers, Vol. I, ILO, Geneva.

Loiseau, I., and Scolnik, H.D. (1976). *Reinterpretation and Adaptation of the Socio-Economic Variables of the Latin American World Model for Studying Different Development Paths Leading to the Fulfillment of the Lima Targets, and Their Effects on the Satisfaction of the Basic Needs,* UNIDO, Vienna.

Mallman, C. (1977). The Bariloche model', Chap. 5 of the book *Problems of World Modelling* (Eds. Karl Deutsch, Bruno Fritsch, Helio Jaguaribe, and Andret Markovitz), Ballinger Publishing Co., Cambridge, Mass.

Scolnik, H.D., Friendlander, A., Loiseau, I., Martinez, M., and Ruiz, C.A. (1977). *Handbook of the Latin American World Model,* UNESCO, Paris.

Fundacion Bariloche (1978). *Complete Technical Reports of the Bariloche Model,* Fundacion Bariloche, Bariloche.

Ruiz, C.A., Loiseau, I., and Scolnik, H.D. (1978). 'Adaptation of the Bariloche model to a national scenario (Brazil)', *Proceedings of the Meeting of Experts on the Applicability of Global Modeling Techniques to Integrated Planning in Developing Countries,* University of Sussex—UNESCO.

The MOIRA Model

The MOIRA model, named for the Greek goddess of fate, is a model of international relations in agriculture. Here is the story of its origins and early history, as described by the project team (Linnemann and coworkers):

> The starting point of the research was the question raised by the Club of Rome: is it possible to produce enough for everyone's needs, even if the world population were to double its numbers? Man's most basic needs in the material sphere may be brought together under four headings: food, shelter, health, and access to education. The initial request of the Club of Rome to some Dutch scholars to analyze the long-term prospects for the world population specified these four basic needs, to be studied in their interrelations and in the setting of the world models as developed by several groups sponsored by the Club of Rome.
>
> In the course of 1972 and 1973, a research group was formed in which a variety of disciplines were represented. It soon became evident that financial and manpower resources available did not permit a sound and thorough study of the various basic needs simultaneously. The group decided to focus its research on one of those needs in particular, i.e., on food. Consequently, the larger part of the group concentrated its research efforts on the international food situation and developed the model MOIRA. Some of the group members continued to work on the broader initial task, without engaging themselves in model building or similar quantitative work.
>
> The project ... was called *Food for a Doubling World Population*; it maintained a rather loose relationship to the Club of Rome as its original inspirer. The moral support received from Dr. Aurelio Peccei and his colleagues of the Club of Rome was matched by the financial support given by various institutions. Research funds were made available by the Ministry of Agriculture and Fisheries and the Ministry of Foreign Affairs in the Hague and by the Haak Bastiaan Kuneman Foundation. Also, several group members were made available to the project by the governing bodies of those institutions where they hold permanent posts. In addition, the Free University at Amsterdam provided office and administrative facilities to the team. (Linnemann *et al.*, 1979, p. vii)

The MOIRA project was conducted in two places. A group composed primarily of economists at the Free University of Amsterdam constructed the formal mathematical model. A group of agronomists at the Agricultural University of Wageningen estimated global limits to food production and consulted on the agricultural production functions in the model.

MOIRA was presented at IIASA in September of 1975. A book describing the project in readable but not sensational terms (Linnemann *et al.*, 1979) was published. Although highly regarded for its clarity and scientific quality it has not received wide distribution.

One aspect of the project, however, did receive considerable media attention, providing further evidence that the limits to growth debate had not yet been laid to rest. The team at Wageningen had been asked to 'compute the absolute maximum of food production of the world, the upper limit of what can be grown on all suitable agricultural land'. In a detailed, carefully argued paper, they presented their conclusion that 'according to the natural restrictions to the growth of agricultural crops, the earth is capable of producing 30 times the present amount of food' (Buringh, Van Heemst, and Staring 1975). This calculation was based on estimates of soil, water, and solar energy only, and assumed no limitations, economic or otherwise, on fertilizers, fossil fuels, labour, or agronomic skills, no pest damage or spoilage, and no use of arable land for any purpose other than agriculture. The team qualified these conditions as neither probable nor desirable. However, media attention focused on the potential for a thirtyfold increase, not on the qualifications, and not on the project's overall conclusions that continuing present trends would increase starvation in the world threefold and that there are major social and political obstacles to meeting the food needs of a doubling world population.

Purposes of the model

Although MOIRA is usually defined as a project focusing on food, it might better be termed a project focusing on hunger. The general issues that motived the research are described by the authors as follows:

> ... the historical record suggests that hunger and malnutrition have remained with us throughout the decades, affecting more and more people as time passes. Although the scourge cf mass starvation and famines has not recurred in the years since the last World War, it is increasingly felt that one cannot continue to cope with food shortages through international trade and aid, but that countries have to rely more heavily on their own capacity to increase food production. But are there ways open to them which promise a better balance between people and their nutrition? Or is the world in its entirety coming dangerously close to the limits of its capacity to produce food? (Linnemann *et al.*, 1979, p. 4)

These goals were the most important for the project:

(1) to describe the world food situation in terms of its underlying causal factors (p. 5).

(2) to provide considered judgements regarding the policy measures that may redirect future developments towards improvements of the world food situation, with special emphasis on international policy measures (p. 5).

(3) ... to try and compute the absolute maximum food production of the world, the upper limit of what can be grown on all suitable agricultural land (see above) (p. 19).

Later, in materials prepared for the Sixth Global Modeling Conference (see Chapter 4 of this book), the group identified two major questions they had analysed by using the model:

(1) What can be said about the development of food deficits in poor countries in the next four decades?

(2) To what extent and in what manner do the agricultural policies of the rich countries affect the development of food production and food consumption in the Third World?

Because the MOIRA model is technically complex and the team members have a low-key style, observers sometimes forget the very strong concerns with world hunger and basic human needs that motivated this project. However, the model reflects these concerns clearly, as do the conclusions and recommendations—and the project team's continuing work.

Structure of the model

Although MOIRA addresses only a subset of the problems investigated by most global models, it is one of the most disaggregated, mathematically elegant, and complex. It identifies 106 nations (3 'centrally planned' and 103 'free market'), and within each nation there are two consumption sectors and one production sector. There are also several 'handles' for experimenting with alternative governmental policies, a principal concern of the model. On the other hand, it distinguishes only one agricultural product, an aggregate called 'consumable protein'. No environmental or energy factors are included at all.

The MOIRA assumptions are based on neoclassical economic theory. Econometric techniques are also used extensively; finding values for externally defined parameters by using statistical estimation procedures is an important concern of the modellers.

Both the supply of and demand for the agricultural product (consumable protein) are included in the model at the national and world levels. It is assumed, with minor exceptions, that the world market will clear at the end of each model cycle (one year). The 'general equilibrium' approach sets world prices at a level where the amount that producer nations export is equal to the amount that consumer nations import. Modest amounts of 'buffer stocks' are assumed (and more as a result of deliberate policy). At the national level, governments are assumed to interfere with the domestic market in order to

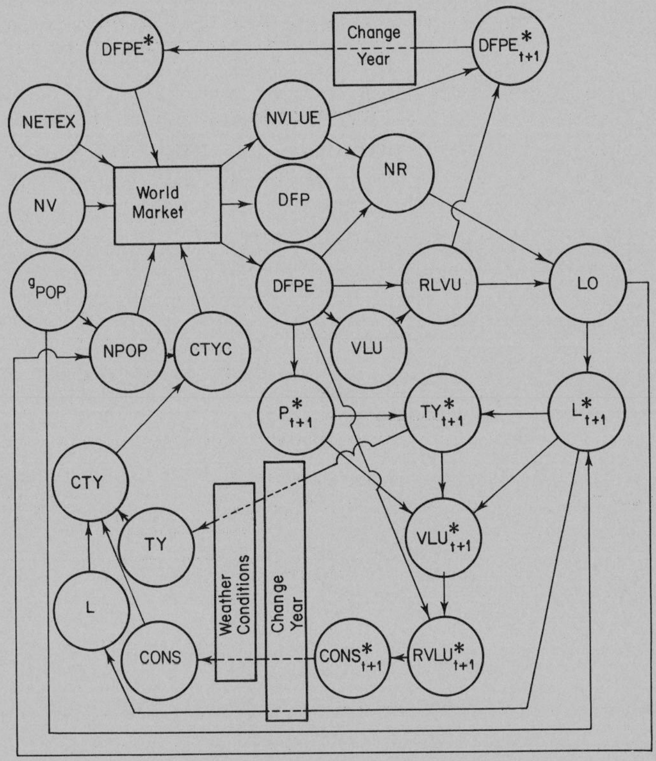

EXPLANATION OF SYMBOLS

CONS = food consumption in the agricultural sector, per capita
CTY = total supply of food by agriculture to non-agriculture
CTYC = supply of food by agriculture to non-agriculture, per caput of non-agricultural population
DFP = world market price level of unprocessed food, measured as deviation from the base-year value
DFPE = domestic price level of unprocessed food, measured as deviation from the base-year value
DFPE* = desired DFPE
gPOP = population growth rate
L = agricultural population
LO = labour outflow from agriculture
NETEX = net food exports of centrally-planned economics
NPOP = non-agricultural population
NR = disposable non-agricultural income per capita, in real terms
NV = (nominal) non-agricultural income
NVLUE = disposable (nominal) non-agricultural income per capita
P = price received by the agricultural producer
RVLU = agricultural income per capita, in real terms
TY = total yield of agricultural production
VLU = (nominal) agricultural income per capita

Asterisks* denote expected values. The time subscript t is suppressed. A horizontal line above the label of the variables indicates an equilibrium value.

Figure 13. The structure of the MOIRA Model. (Reproduced by permission of Elsevier North Holland Inc. from Linneman *et al.*, 1979, p. 283)

meet national goals such as price stability and balance-of-payments needs. Policies of national governments include regulation of domestic food prices, trade, and foreign assistance.

The heart of MOIRA is the mathematical algorithm (that is step-by-step problem-solving procedure) for optimizing production decisions within the national agricultural sectors and for balancing production and demand (clearing the market) at the international level. No diagram or flow chart can convey a complete picture of this complex process; however, we have included Figures 13 and 14 to suggest the elements of the analysis.

Figure 13 shows some of the interrelations among variables in the MOIRA model. The box in the upper left labelled 'world market' is a good place to begin looking at the diagram. The inputs to the world market from each national model are:

- (nominal) non-agricultural income (NV),
- net food exports of centrally planned economies (NETEX),
- non-agricultural population (NPOP),
- supply of food by agriculture to non-agriculture as per capita of non-agricultural population (CTYC),
- desired domestic food price (DFPE*).

Since there is no non-agricultural production sector, non-agricultural income must be supplied as an external value. The remainder of the values are computed by the national models.

The outputs of the world market are:

- disposable (nominal) non-agricultural income per capita (NR),
- world market price level of unprocessed food measured as a deviation from the base-year value (DFP),
- domestic price level of unprocessed food measured as a deviation from the base-year value (DFPE).

Figure 14 presents a somewhat different picture. It shows the computation sequence of the model and depicts somewhat more clearly the major processes included. It is worth noting, in particular, that the computations relating to the world market appear midway in the sequence. This means that the world market both affects and is affected by the behaviour of individual countries in each model cycle.

With the possible exception of SARUM, the analysis of demand for agricultural products in MOIRA is by far the best in any global model. Population growth is based on externally defined variables. Two regions (urban and rural) and six income classes are identified in each nation, for a total of twelve population categories. This means that food demand can be estimated for each of twelve different groups in each country. The

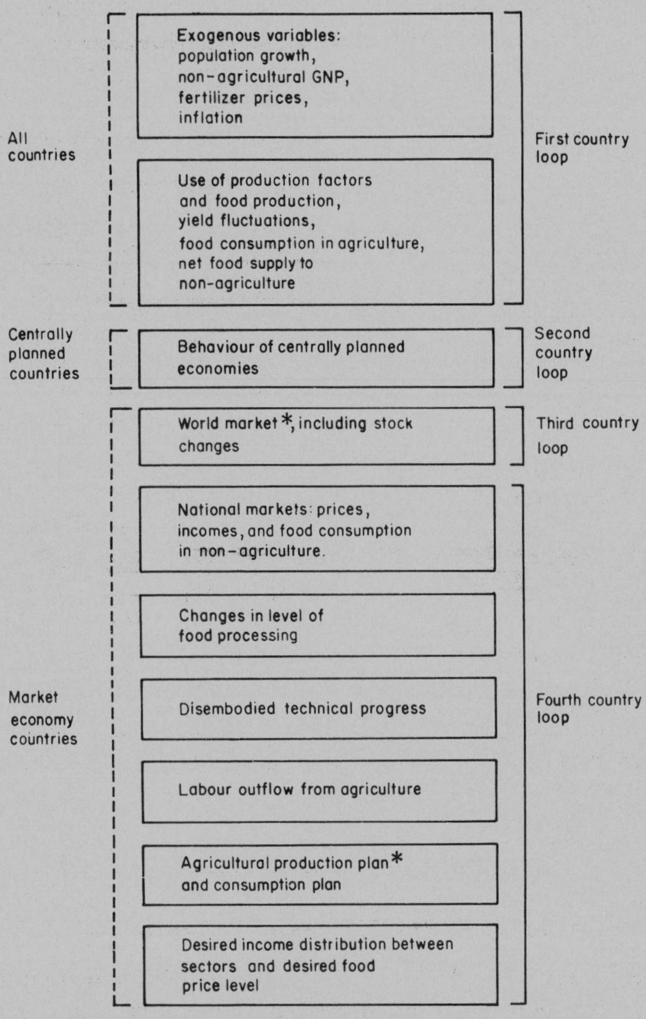

Figure 14. The general structure of the MOIRA computer program. (Reproduced by permission of Elsevier North Holland, Inc. from Linneman, 1979, p. 284)

mathematical basis for making these estimates is presented with considerable care.

The submodel that represents the production sector assumes that producers decide in each yearly cycle to optimize their production, based on prices, factor costs, and technological considerations reflecting regional differences

Table 3. A selection of results from the standard runs of the MOIRA model (from Linnemann *et al.*, 1979, p. 306)

	1966	1975	1980	1990	2000	2009
Total agricultural production (10^6 kg consumable protein)						
World	1943	2495	2970	4094	5340	6530
Developed countries	743	910	1067	1517	1896	2194
North America	403	497	580	855	1009	1151
European Community	147	182	218	297	377	434
Developing countries	599	749	941	1237	1774	2344
Latin America	181	241	297	434	631	905
Tropical Africa	66	82	96	145	208	297
Middle East	58	72	96	128	220	287
Southern Asia	265	321	409	467	624	736
Food consumption per capita (kg consumable protein)						
World	46	50	53	61	68	73
Developed countries	86	102	110	130	147	160
North America	107	121	129	148	163	173
European Community	90	103	108	126	141	153
Developing countries	30	30	33	37	43	48
Latin America	43	48	52	63	76	89
Tropical Africa	25	25	27	34	40	47
Middle East	38	40	46	56	71	78
Southern Asia	26	25	26	27	29	29

and past experiences. With this structure, the model can incorporate the view that government policies do not affect production directly, but, rather, affect the motivations of producers. This model structure also emphasizes the fact that in today's world—and in the foreseeable future—it is poverty, much more than supply constraints, that is the cause of world hunger. People are not hungry because there is not enough food; they will not starve because there is insufficient potential to grow food, at least in the foreseeable future. They are hungry and they will starve because they cannot purchase the food that is there.

Results

Among world models MOIRA is one of the most complete and clear in presenting its results. We divide our summary into four parts:

	1966	1975	1980	1990	2000	2009
Self-sufficiency ratio of consumable protein						
Developed countries	96	92	97	107	110	109
North America	142	143	149	174	175	176
European Community	59	60	67	74	81	82
Developing countries	106	102	104	93	93	91
Latin America	123	113	110	103	100	101
Tropical Africa	111	104	102	98	93	92
Middle East	87	85	88	58	65	65
Southern Asia	100	101	108	95	97	95
Population in agriculture (%)						
World	52	46	42	36	32	29
Developing countries	66	60	57	53	48	45
Latin America	45	38	36	32	28	25
Southern Asia	70	64	61	57	52	48
Price indicator world food market (DFP: 1965 = 0)	– 8	60	431	301	422	442
World hunger						
Total food deficit (WHUNG) (10^6 kg consumable protein)	15	29	40	43	62	97
Million people below minimum food standard	180	350	480	520	740	1160
WHUNG located in agricultural sector (%)	76	85	35	42	19	15

1. The standard run
2. Sensitivity-analysis runs
3. Analysis of international food policies
4. General conclusions

1. *The standard run*

As in the case of most models, the term 'standard run' is used to denote an unfolding of events that is essentially a continuation of present trends. In particular, this means that the values of exogenous variables are based on extrapolations from the past and no new major policy interventions are assumed. The principal externally supplied inputs are population growth, growth in production of non-agricultural goods and services (GDP), fertilizer prices, and nutrition standards. Table 3 shows a selection of tabulated results from the standard run.

A good place to begin looking at this rather formidable table is at the bottom, with the summary variables *total food deficit* and *million people below minimum food standard.* It should be remembered, also, that the trends across the table from left to right are more important than specific numerical values. In the standard run, the number of people below the minimum food standard in the year 2000 is somewhat more than twice the number in 1980, despite the fact that world food production will have nearly tripled. Tropical Africa and Southern Asia remain the major food-deficit regions. The trend in developing nations is toward decreasing, rather than increasing, self-sufficiency in food. By the year 2000 North America will have strengthened its position as the dominant exporter of basic foods.

2. *Sensitivity-analysis runs*

In these runs, the modellers examined changes in these exogenous assumptions:

- the impact of economic growth outside of agriculture,
- the impacts of population growth,
- the responses to changes in income distribution outside agriculture.

The reported runs represent only a small fraction of those actually undertaken with the model.

- *The impact of economic growth outside of agriculture.* According to the model, developments in the world situation are rather sensitive to the growth rate of non-agricultural GDP. For example, a reduction of the economic growth rate by half generates much lower market prices in the agricultural sector and increases world hunger by 35 per cent.

- *The impacts of population growth.* 'In order to test the sensitivity of the model to changes in the growth rate of population, a simulation run was made in which the growth rates were set at half their values in the standard run.' The results include an increase in income outside agriculture, a reduction in the agricultural labour supply and, concomitantly, a decline in total demand. On the average, hunger is 30 per cent less than in the standard run. The authors conclude '. . . that the much slower rate of population growth which was assumed alleviates the world food problem but does not solve it' (Linnemann *et al.*, 1979, p. 298).

- *Sensitivity to income distribution outside agriculture.* In this run, income inequality in the non-agricultural sector is gradually reduced to about half its initial magnitude during the 1975-2010 period. 'This simulation run illustrates the thesis that the hunger problem is, to a large extent, a problem of income distribution' (Linnemann *et al.*, 1979, p. 300). The results are substantially lower food deficits and a reduction by half in the extent of world hunger. Unfortunately, the model structure does not allow the assumption of greater income inequality in the agricultural sector to be examined.

In concluding the discussion of their sensitivity analysis, the authors emphasize the limitations of the model. But they point out that 'all simulation runs with alternative assumptions regarding exogenous variables have one thing in common: if policies remain unchanged, the number of people who cannot obtain sufficient food will increase' (Linnemann *et al.*, 1979, p. 303).

3. *Analysis of international food policies*
In the final chapter of their report, the MOIRA analysts examine policies designed to alleviate world hunger, investigating measures designed to redistribute food supplies and to stimulate food production.

- *Redistributive measures.* The first redistribution measure they examine is the reduction of food consumption in the rich countries (a step most likely to be achieved by shifting consumption patterns, such as reducing the amount of animal products consumed). The result is a low world market price that has an adverse effect on food production, resulting in no improvement in food supply in the countries where hunger prevails.

  A more successful programme involves the rich countries financing food purchases by the poor in an amount equal to about 0.5 per cent of each rich country GNP. This scenario does eliminate hunger.

  The most significant difference between these two redistributive measures is that the latter one acts as a stimulant to food production in the agricultural sector.

- *Stimulating food production in the developing countries through international action.* The policy of stabilizing world food prices at relatively high levels proves to have a significant impact on hunger. The principal effect is to induce increased production, which also has beneficial spillover effects on consumption, especially in the agricultural sector. This stabilization policy could be implemented, according to the model, if the rich countries would establish buffering stockpiles sufficient to bridge the yearly fluctuations in supply and demand. In addition, it is assumed that the rich nations adopt an export-import policy that will support the internationally-agreed-upon world price, which requires domestic policies to influence consumption, production, or both, and probably means greater internal food price fluctuations in the rich countries.

An alternative policy, liberalizing food trade, turns out to have an adverse effect on world hunger. This policy lowers world prices, lowers production in the developing nations, and increases hunger. It benefits the rich countries at the expense of those less fortunate.

Overall, the MOIRA team concludes that a programme to alleviate world hunger must combine market stabilization with an aid programme designed to stimulate demand, and that this policy must be initiated and pursued by the rich nations of the world.

General conclusions

Although the MOIRA project focuses on food and hunger, its general conclusions echo themes that have been emphasized in other global modelling results:

> The international policy for agriculture and food supply advocated above requires that the rich countries significantly reorient their agricultural policies. The rivalry between the economic blocs within the western world (i.e., between North America and the Common Market) with regard to agricultural policies will have to be replaced by a concerted effort towards a global food supply policy. In pursuing their national agricultural policies, the rich countries will have to make allowances for the targets of this international policy. The desire to stabilize world market prices at a fairly high level entails that the rich nations should make more of an effort to adapt the volume of their own food production (and, if necessary also their own consumption) to the international demand and supply situation. ... With this purpose in mind, the rich countries will have to extend and complement the instruments of their agricultural policies so that both national and international targets can be realized.
>
> The conditions under which regulation of the world market would be possible are not likely to be fulfilled from one day to the next. All too frequently, in fact, international cooperation founders on the priority of national interests. It is to be hoped that the MOIRA study will be able to support the efforts to put international economic relations at the service of a target transcending narrow national interests, in this case, sufficient food for all. (Bruckmann, 1975, p. 38).

Further developments

Like several other projects, MOIRA was, in its early stages, a 'shoestring' operation, which survived on the willingness of the team members to donate their time and effort. It was not until 1975, the year the model was presented at IIASA, that available funds began to approach the level of effort expended. Since then, however, support has improved steadily.

In July 1976, a foundation was established to continue the MOIRA work. Two ministries of the Netherlands Government (Agriculture and Development Cooperation) and two universities (the Free University in Amsterdam and the

Agricultural University in Wageningen) participate. Wouter Tims, formerly of the World Bank, has been named Director of the new Center for World Food Studies.

Although much of the team's efforts now focus on more narrowly defined problems, the global perspective of the original project has not been lost. It is retained through close collaboration between the group in the Netherlands and IIASA's Food and Agricultural Project. Michiel Keyzer, who developed MOIRA's international trade mechanism, has adapted this part of the model to the needs of the more ambitious food model at IIASA; he collaborates with the IIASA team on a regular basis. The continuing work, both in the Netherlands and at IIASA aims to contribute further to our knowledge of what causes hunger in the world and how it can be eliminated.

The principal references

Linnemann, H., De Hoogh, J., Keyzer, M., and Van Heemst, H. (1979). *MOIRA- Model of International Relations in Agriculture,* North-Holland Publishing Co. Amsterdam.

Prior to the publication of this volume, the framework for analysis and a number of preliminary findings have been published in several articles and research reports. These forerunners of the main report, in English, were, in chronological order:

Population Doubling and Food Supply (1974). Economic and Social Institute, Free University, Amsterdam. July 1974.

'Problems of population doubling: the world food problem' (1974). In: *Latin American World Model- Proceedings of the Second IIASA Symposium on Global Modeling* (Ed. Gerhart Bruckman), October 1974, pp. 305-310, IISAS, Laxenburg, Austria.

Buringh, P., Van Heemst, H.D.J., and Staring, G.J. (1975) *Computation of the Absolute Maximum Food Production of the World,* Agricultural University, Wageningen, January 1975.

Food for a Doubling World Population (1975). Report on the Club of Rome Conference organized by the Austrian College Society at Alpbach, Tyrol, Austria, 25-27 June 1975. 'Food for a growing world population' (1975). In *MOIRA: Food and Agricultural Model- Proceedings of the Third IIASA Symposium on Global Modeling* (Ed. Gerhart Bruckman), 22-25 September 1975, pp. 13-515, CP-77-1, IIASA, Laxenburg, Austria.

'Food for a growing world population: some of the main findings of a study on the long term prospects of the world food situation' (1976). In *European Review of Agricultural Economics,* **3**, 459-499.

Garbutt, J., Linnemann, H., *et al.* (1976). *Mensen tellen,* Chap. 6, Spectrum, Utrecht.

'Food for a growing world population' (1977). In *Technological Forecasting and Social Change,* **10**, 27-51.

Linnemann, H., and De Hoogh, J. (1977) 'Probleme der Welternahrung und der Neuen Weltwirtschaftsordnung', in *Zeitschrift fur Evangelische Ethik,* **21**, 136-143.

Linnemann, H., *et al.* (1977). 'Prospettive in Agricolture: MOIRA, uno Studio Mediante Modello', in *Enciclopedia della Scienza e della Tecnica,* Ed. Mondadori, pp. 113-124.

The SARU Model

The mission of the Systems Analysis Research Unit (SARU), located in the UK Department of the Environment and headed by experimental physicist Peter C. Roberts, was to 'explore the implications for national and international policies of long-term trends in environment and society' (SARU Staff, 1977, p. viii). One of the ways they filled this mission was to inform themselves about the global models under construction, their methods, data bases, and major assumptions.

Roberts and his team quickly gained respect as constructive critics in the emerging global modelling field. At IIASA conferences, other professional meetings, and in several publications, the SARU staff has provided informed, solid commentary on the principal models. Speaking from an 'in-house' perspective, Roberts places particular emphasis on the needs and special requirements of policy-level officials. 'How could such models be made more credible?' he asks. 'At what point in their development should we expect models to make a really effective contribution to the decision-making process?'

In probing these questions, the SARU staff 'came to the conclusion that the difficulties could best be understood by attempting to build a world model within the unit' (SARU Staff, 1977, p. viii). The result of their efforts, 'a simulation model making use of standard rules of economic behavior and based, as far as possible, on published empirical data', called SARUM, was first presented at IIASA in September 1976.

The SARU model and the organization of the team have several particularly interesting features. First, the SARU staff self-consciously had limited aspirations for their model. Roberts' skepticism about the utility of pioneering global models extended to his own work. 'At this stage of development,' he has emphasized, 'it is unwise to lay much emphasis on the character of the "futures" which are simulated. Of more importance is establishing the validity of the data base and the relationships. Not until there is some confidence in the soundness of the basic work is it appropriate to draw conclusions and make recommendations.' (SARU Staff, 1977, p. ix).

A second feature of the project is its built-in self-criticism mechanism. The Science Policy Research Unit (SPRU) at Sussex University, which had gained international visibility with its criticism of *The Limits to Growth* (Cole *et al.*, 1973), was *retained* by SARU to play a similar role vis-à-vis their model. The SPRU review was included in the IIASA presentation of SARUM— although not in the subsequent *Research Report* (SARU Staff, 1978) describing the model.

The absence of a popularly oriented publication emphasizing conclusions is a third distinguishing feature of this project. The group did publish excellent documentation, however, and has made versions of the model readily available to anyone wanting to implement and run it.

Finally, SARU learned from and built upon the work of other projects. Although the model reflects a commitment to a structure 'based on relationships implicit in the neoclassical synthesis', SARUM has incorporated ideas from system dynamics, econometrics, input-output analysis, and the multilevel approach used by Mesarovic and Pestel. Where the model attempts to break new ground, the authors are very explicit about what they are doing and why.

Although an advanced, fifteen-region version of SARUM was used by the OECD Interfutures Project (1979) for many of its projections, the model has not received wide media attention or public recognition. In view of the group's philosophy, this is not surprising. But among global modellers, as well as other modelling professionals who are familiar with the model, both the model and the project team are held in very high regard. This group has adhered to high standards of scientific inquiry and good modelling practice. There is much to be learned from SARUM and from the development process that culminated in its completion and documentation.

The purpose of the model

To the query, 'What are your purposes?', the SARU staff would respond:

1. to learn more about the difficulties of building global models and how to overcome them;
2. to strive for a deeper understanding of the way the global system works; for a greater knowledge of the effects of interaction than can be obtained from analysis of selected isolated problems;
3. to discover the possible areas and dimensions of stress of the development of the global system.

In a general way the problem focus of SARUM is similar to that of the other global models. It is concerned with long-term developmental trends in the global system, with the impacts of the policies that may be designed to shape these trends, and with avoiding hazards.

The structure of the SARU model

SARUM is a multiregion, multisector model, somewhat similar in geographical disaggregation to that of Mesarovic and Pestel. The earliest version was highly aggregated, incorporating only three regions (or strata) based on income levels; however, the latest version defines fifteen regions. Regions are linked by trade and aid; both commodity and monetary flows are defined. Each region has an identical structure, comprising thirteen interlinked economic sectors. The documented version of the model devotes nine sectors to a rather disaggregated representation of agricultural activity. Non-agricultural activities are more highly aggregated into four sectors: minerals,

Table 4. The sectors in the SARU model (from SARU staff, 1977, p. 30)

Sector (units)	Short description of commodities produced
Raw grain and starchy roots production (kg)	Grains Fresh potatoes Cassava, sweet potatoes, and yams
All other vegetables and fruit production (kg)	Fresh (or dried) fruits and nuts Pulses, tomatoes, and other fresh vegetables Oil seeds including soya beans and oil palm
Vegetable processing (1968 US dollars*)	Edible grain products Bran — non-edible grain by-products Oil seed cake and meal Other food wastes and prepared animal feeds Oil-seed flour or meal Fixed vegetable oils and fats Margarine Fruit, dried or preserved by canning, freezing, etc. Vegetables, frozen, preserved, or prepared Sugar, honey, and sugar products Miscellaneous prepared foods
Non-food crops (1968 US dollars*)	Coffee, tea, cocoa, and spices Unmanufactured tobacco Non-consumable oil seeds — linseed and castor Cotton, jute, and other vegetable fibres Crude vegetables for dyes, gums, etc. Planting material, flowers, and ornamentals Hay and fodder crops
Agricultural services (kg)	Fertilizers, manufactured (nitrogenous, phosphatic, and potassic) Insecticides, fungicides, sheep and cattle dressing, weed-killing preparations
Livestock production (1968 US dollars*)	Live animals and fresh meat Eggs, fresh milk, and fish

primary energy, capital, and non-food goods and services (see Table 4). Interregional differences are taken into account by defining different initial conditions and parameter values for each region. The principal exogenous inputs to the model are population, technological growth in capital and labour productivity, technological growth in fertilizer response, and a representation of income distribution within the population of each region.

The basic assumptions

A good place to begin discussing the SARUM model is with the list of the basic theoretical assumptions, which are based on neoclassical economics, a tradition that traces its roots to Adam Smith's *Wealth of Nations* (1776), and

Sector (units)	Short description of commodities produced
Livestock processing (1968 US dollars*)	Meat preparations and dairy products Lard and animal oils and fats respectively Fish, fresh or preserved
Non-food products and services (1968 US dollars*)	Beverages, hides, wool, other crude materials, waste paper, and synthetic fibres Manufactured gas, electricity, and petroleum products Non-agricultural chemicals Manufacturers of leather, rubber, wood, paper, textiles, minerals, and glass Telecom apparatus and domestic electrical equipment Passenger motor cars and motorcycles Sanitary plumbing, heating, lighting fittings, furniture, travel goods, clothing, footwear, photographic supplies, clocks, and miscellaneous manufactures Crude rubber Rough wood and timber products
Capital (1968 US dollars*)	Construction Non-electrical machinery Electrical transmission and distribution equipment and industrial apparatus Transport equipment excluding private motor vehicles and motorcycles Scientific and optical instruments
Primary energy (1968 US dollars*)	Coal Crude oil Natural gas
Minerals (1968 US dollars*)	Fertilizer minerals Metal ores and scrap Stone and clay mining and quarrying
Land (ha)	Developed land for agriculture
Water (10 m^3)	Irrigation water for agriculture

*That is 1968 US dollars-worth; a physical quantity measured in constant units of money.

reflects, in particular, the later theoretical contributions of John Maynard Keynes (see *The General Theory of Employment, Interest, and Money*, Harcourt Brace, New York, 1935). Nine basic propositions are identified by the SARU staff (SARU staff, 1977, p. 9):

1. Consumers will choose the goods and services that maximize their utility, given the prices of the goods and services.
2. Entrepreneurs will select the production technique that maximizes their profit, given the factor prices.
3. Producers cannot affect the prices of the goods they sell by adjusting the output.
4. Producers cannot affect the prices of the goods they buy by adjusting demand.

5. Labour and investment will be attracted to these industries where the wages and profits are highest.
6. Prices are an inverse function of stocks.
7. A set of prices exists that will clear all markets, apart from desired stocks.
8. Consumers will buy more (less) of a good as its price falls (rises), unless the good is 'inferior'.
9. Entrepreneurs can enter or leave any industry, depending on whether they make a profit or loss.

The overall goal incorporated in the model is to improve the material well-being of deficient members of the population without making anyone worse off (Pareto optimality).

The modellers recognize that these assumptions are not universally accepted and may be regarded as naive by some. However, they argue that the usefulness of these assumptions, as applied in the model, is supported by empirical data. The detailed arguments are found in SARUM 76, Chapter 3, and make interesting reading for both supporters and opponents of this perspective.

Application of the assumptions: the structure of a basic sector

To grasp the way in which the 'neoclassical synthesis' has been incorporated in SARUM, it is useful to look more closely at the model's representation of an economic sector. Figure 15 shows the basic structure. (Not every sector conforms precisely to this representation—some agricultural sectors, in particular, are a bit different—but the basic approach is similar throughout.) This structure is also of interest because it shows similarities between the SARU approach to modelling and system dynamics. The dominant feedback loops in the model determine labour supply and wages, production levels (output), capital accumulation, and prices. The price mechanism incorporates time delays (smoothed demand and smoothed cost) to take into account the delayed response of prices to changing costs and patterns of demand. Classical Cobb-Douglas production functions are used to relate labour and capital to output. Non monetarized and monetarized sectors are linked. Computations establish the values of the different variables, and especially the balance (equilibrium) between supply and demand. Sectors within each region are linked by flows of labour, capital, and commodities.

The linkages among sectors

Figure 16 shows these linkages. The between-sector flows are determined by prices and profitability. Wage rates determine the flows of labour from sector to sector.

The linkages among regions

Five assumptions about international trade are made in the SARU model:

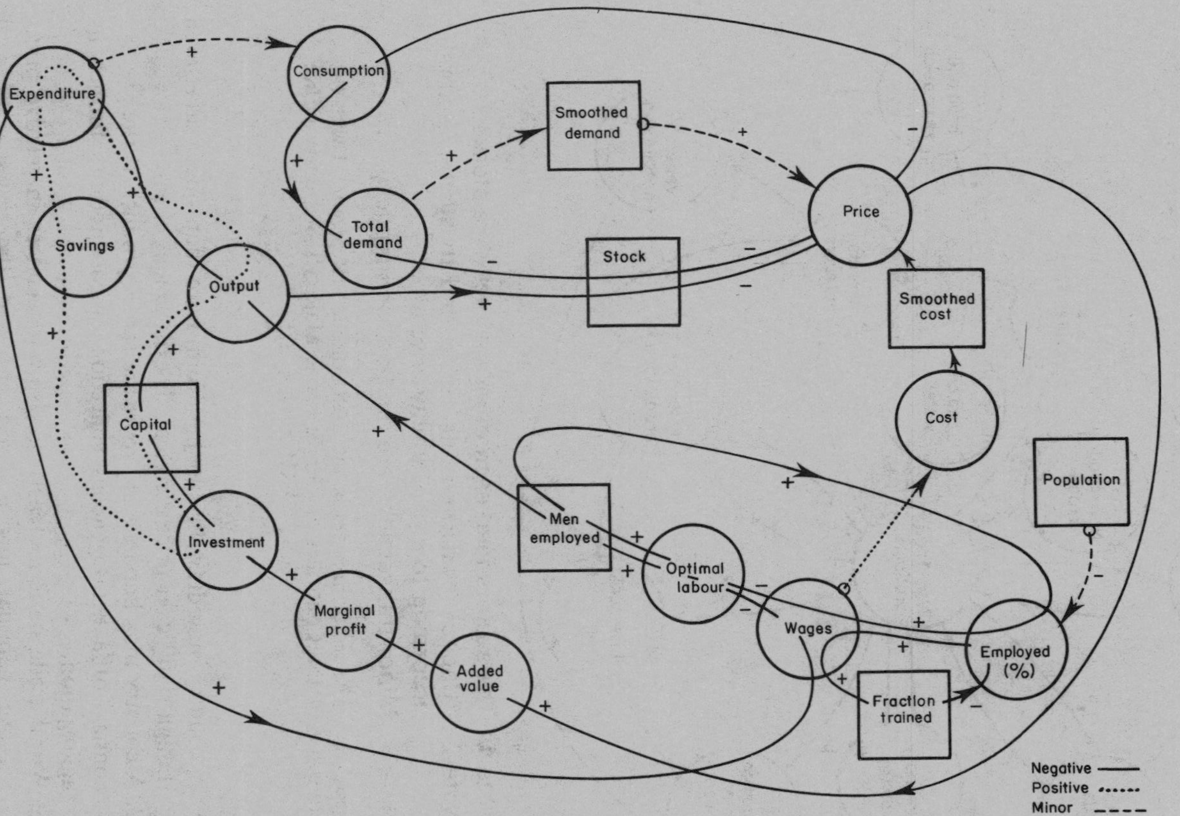

Figure 15. The principal variables and dominant feedback loops in the sectoral structure of SARUM. (Reproduced with the permission of the Controller of Her Britannic Majesty's Stationery Office from SARU staff, 1977, p. 23)

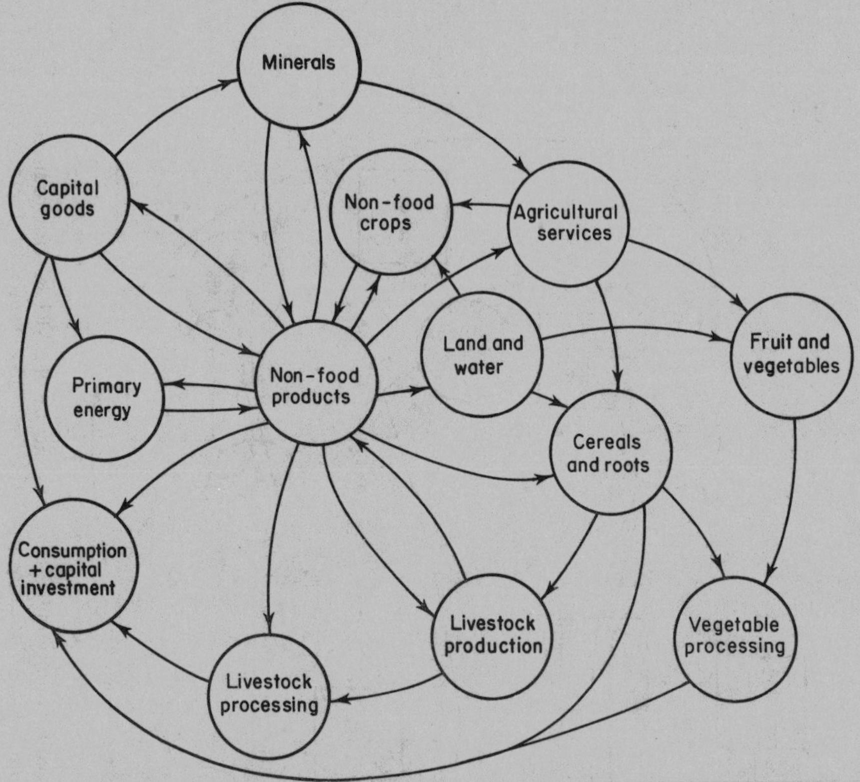

Note 1 The sectors feed each other indirectly via stocks.
Note 2 "Capital investment" is distributed to all the sectors according to a profitability criterion.
Note 3 Minor transactions are omitted for simplicity.

Figure 16. The sector linkages in the SARU model. (Reproduced with the permission of the Controller of Her Britannic Majesty's Stationery Office from SARU staff, p. 27)

1. A country may draw its supplies from more than one source, even though significant price differentials exist among them.
2. A country may import and export the same 'product' if the product consists of a heterogeneous collection of commodities formed by aggregation.
3. World trade patterns react sluggishly to changes in relative prices because habitual business relations and the massive physical infrastructure (railways, docks, pipelines, etc.) do not change quickly.
4. Some countries regard trade regulations and tariffs as an extension of strategic and diplomatic policy and use trade to further their interests abroad and protect special interests at home.

5. World trade calls for shipping goods over greater distances than domestic trade, and the cost of transport may be a decisive element in the total cost at the market.

A simple 'bias matrix' is used by the modellers to take into account deviations from normal price-determined behaviour. Numbers in the matrix correspond to each region and each sector and measure the degree of deviation. This permits the aggregate effects of factors such as distance and tariff barriers to be incorporated. Nations may run trade deficits for a period of time. Heavy debts lead to reductions in domestic expenditures and adjustments in the trade bias matrix to reflect a lack of international credit.

The final international flow, foreign aid, is represented in this version of the model as a simple flow of money. One regions's expenditure is increased and another's decreased by a specified proportion of the donor's gross regional product.

Results

Because the SARU staff has argued so strongly against drawing unwarranted conclusions from model results, we will not dwell upon this topic in our discussion of their model. One illustrative result, focusing on agriculture, will suffice; Figure 17 arises from a scenario in which there is a rapid rise in food

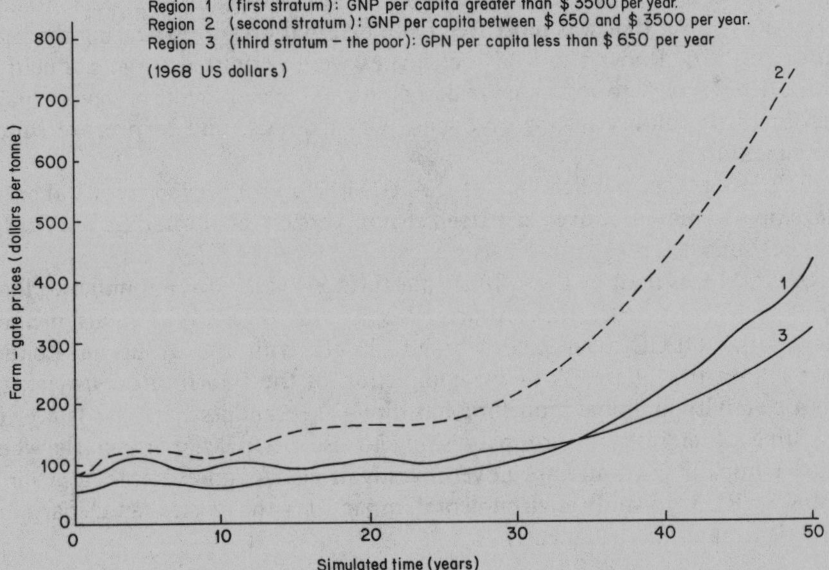

Figure 17. The results of a SARUM 'experiment': the rises in cereal prices due to population pressure and land scarcity. (Reproduced by permission of the Controller of Her Britannic Majesty's Stationery Office from SARU staff, 1977, p. 201)

prices due to population pressure and scarcity of land. The modellers comment on this 'experiment' in this way:

> The emphasis of the first experiment is on food production because, in resource terms, the problem of feeding a growing population is more acute than either supplying it with adequate energy or of maintaining sufficient mineral inputs. Two depletion effects dominate the further expansion of agriculture. It becomes increasingly costly to develop each hectare of currently unused land to the point where it rises as the flow increases. The depletion is partially offset by the higher productivity of labor, not only in agriculture but also in the sectors ancillary to agriculture (machinery, processing, and distribution). The rates of technical change observed in the past have been used in the initial growth phase. However, perpetually improving technical change has been eschewed, both because the concept of potentially infinite improvement is intrinsically improbable, and because critics might argue that the results depended crucially upon this unlimited improvement. (SARU staff, 1977, p. 195).

In general, the results of the SARUM 'experiments' do not seem to point in a direction significantly different from those of other models discussed here. Rather, they appear, for the most part, to support the broad patterns that these models have identified, albeit in more circumspect and muted tones.

Further developments

SARUM is the only global model among those reviewed that was funded by a government agency and developed in-house. Paradoxically, it appears to have been the project that was most free from external pressures. During the years prior to 1976, Roberts and his colleagues were permitted to go about their work deliberately, to follow an orderly path of development, to devote major resources to validity testing and sensitivity analysis, and to prepare superb documentation.

Following the publication of SARUM 76, a decision was taken to develop— as noted above— a fifteen-region version of the model and to 'go public' with it.

SARUM was used by the OECD Interfutures Project for a number of long-term scenarios (OECD Interfutures Project, 1979). A SARU staff member joined the OECD team and worked closely with the group in London throughout this phase. With the completion of the Interfutures Project, the Unit's activity in global modelling has diminished and its emphasis is now on assisting other research groups wishing to use SARUM, particularly where such groups plan to attempt novel investigations. As an example, a group is using SARUM to study environmental impacts for the Australian Department of Science and Environment.

The principal references

Cole, H.S.P., Freeman, C., Jahoda, M., and Pavitt, K.L.R. (1973). *Models of Doom,* Universe Books, New York.

Cottrell, A. (1973). 'Problems of predicting future world trends', *Nature,* **245**,280-283.
Roberts, P.C. (1973). 'Models of the future', *Omega,* **1**,No. 5, 591-601.
Clark, J., and Cole, S. (1975) *Global Simulation Models,* J. Wiley & Sons, New York.
Norse, D. (1975). 'Development strategies and the world food problem', Agricultural Economics Society, 1975 Conference.
Norse, D., and Roberts, P.C. (1975). 'The problems of food production in certain Asian countries', Third IIASA Global Modeling Symposium, Vienna, Austria.
Roberts, P.C. (1975). 'The world can yet be saved', *New Scientist,* 23 January **1975,** 200-201.
Roberts, P.C. (1975. 'Large scale systems modelling', First European Congress on Operations Research Conference, Brussels, January 1975.
Cabinet Office (1976). *Future World Trends—A Discussion Paper,* HMSO, London.
Roberts, P.C. (1976) 'Models for prediction under conditions of interaction', NATO Conference on Environmental Assessment of Socio-Economic Systems, Istanbul, Turkey, October 1976.
Mellanby, K. (1977). 'Model report', *Nature,* **269,** 555.
Parker, K.T. (1977). 'Modelling intra and inter regional activity', Second International Conference on Dynamic Modelling and Control of National Economics, Palais Auersperg, Vienna, January 1977.
Roberts, P.C. (1977). 'SARUM 76—a global modelling project, *Futures,* February **1977,** 3-16.
Roberts, P.C. (1977). 'Adaptive mechanisms in global models', Fifth IIASA Global Modeling Conference, Vienna, 1977.
Roberts, P.C. (1977). 'Futures research in government', British Association Conference Aston, September 1977.
SARU staff (1977). 'SARUM 76 — global modelling project', Research Report No. 19, UK Departments of Environment and Transport, London.
Parker, K.T., and Raftery, J. (1978). 'The SARUM global model and its application to problems of interest to developing countries', International Conference on Systems Modelling in Developing Countries, Asian Institute of Technology, Bangkok, May 1978.
Roberts P.C. (1978). 'Learning processes in global models, Sixth IIASA Global Modeling Conference, Vienna, 1978.
Roberts, P.C. (1978). 'Teleological mechanisms — the concept of anticipation', *Proceedings of the Second International Seminar on Trends in Mathematical Modelling,* Jablonna near Warsaw, December 1974; Ossolineum pp. 233-241.
Roberts, P.C. (1978). *Modelling Large Systems,* Taylor and Francis, London.
SARU Staff (1978). *SARUM Handbook,* SARU Departments of Environment and Transport, London.
OECD Interfutures Project (1979). *Facing the Future: Mastering the Probable and Managing the Unpredictable,* OECD, Paris.

The FUGI Model

We have traced the impact of the global *problematique* as defined by the Club of Rome through the Forrester/Meadows models, reactions to them by Mesarovic/Pestel and Bariloche, elaboration of the food sector by the MOIRA group, and the refined commentary on the whole that led to SARUM. The statement of the *problematique* also produced reverberations in Japan, at first through the interest of Saburo Okita, an influential government advisor and a member of the Club of Rome Executive Committee. The first attempts at global modelling in Japan were conducted by Professor Yoichi Kaya, a systems scientist of the University of Tokyo and were presented to the Club of Rome at their 1974 Tokyo meeting.

FUGI is the only global model to be made by a non-Western modelling team. Like Bariloche, it was motivated by concerns about the harmonious development of Third World nations, together with continuous progress of the industrialized areas. It shares with SARUM a degree of eclecticism in its approach. Like the Mesarovic Pestel and MOIRA models, it has been produced by collaborative efforts of several geographically separated groups.

FUGI was the first of the global models presented at IIASA to rely heavily on a well-established economic modelling technique—input-output analysis. (SARUM was based on *economic theory,* you will recall, but applied in a relatively unconventional manner.) A major contribution of this project is its approach to overcoming the essentially static, short-term character of input-output analysis by coupling it with a dynamic macroeconomic model. To model long-term trends in the energy sector, a system dynamics model was used.

No popularly oriented book has yet been written by the team; nor has the model received much popular attention. However, it is widely discussed by professional economists and, presumably, is well known in Japan. Since 1977, FUGI has undergone considerable refinement and development. Officials at UN-ESCAP (Economics and Social Commission of Asia and the Pacific) have taken an interest in the model, and the present version places a particular emphasis on the problems of the Pacific area.

The purposes of the model

Among the global modelling groups, the FUGI team is one of the most eloquent in its statement of broad goals, specific objectives, and a vision of a new global order.

The modellers begin with an expression of concern about the capability of economic sciences to address emerging global problems:

> ... the frameworks of the economic sciences in the past are losing their efficacy and becoming unable to respond to many of the changes inherent in the new conditions that typify the world scene.
>
> This is in part because the angles of vision utilized by economists in the past have been too narrow .... Efforts to look squarely at the realities of [the] interdependence among countries and to systematize them within a framework of economic theory have up to now remained, by and large, outside the purview of economists. (Kaya, Onishi, Suzuki, *et al.*, 1977).

Like other groups whose models were motivated by the problematique of the Club of Rome, the FUGI staff hoped to develop scenarios and policy recommendations that would contribute to solving global problems. The most serious problems, in their judgement, involve future trends of industrialization in developed and underdeveloped nations as they will affect an increasingly interdependent world economic system. Obviously the future of Japan as an industrialized nation dependent upon a large export market is a matter of specific concern. The authors elaborate as follows:

> The most serious among a number of problems concerned with industrialization is the competition in the world market of manufactured goods between advanced industrialized countries and developing ones. The former countries have so far dominated the world market, and the countries with scarce resources such as Japan have to live still on the export of these goods. Nevertheless, some of the latter countries are very poor in resources except human resources, so that their future heavily relies on development of manufacturing industry .... Will the harmonious growth of these two kinds of countries be possible? If possible, in what way? (Kaya, Onishi, Suzuki, *et al.*, 1977, p. 96)

In pursuing the answer to this question, the authors decided they needed a relatively detailed analysis of the interactions among industrial sectors in regional and national economies. This led them to use the approach customarily employed by economists to examine interindustry interactions—input-output analysis. But input-output analysis produces results only for a single year; it is not dynamic and it is principally useful for short-term projections. Because of this, the FUGI staff developed a more highly aggregated dynamic model, which could track the most important economic trends over the long run and provide necessary inputs to the disaggregated input-output model for the specific years of interest. For the analysis of problems relating to materials and energy, a third approach, system dynamics, was used. However, it does not appear that the system dynamics model has been linked directly to either the macroeconomic or input-output model.

The structure of the model

This section will describe briefly the three major components of the FUGI model:

1. The global input-output model (developed by the group led by Yoichi Kaya at Tokyo University).
2. The global macroeconomic model (developed by Akira Onishi and his team at Soka University).
3. The global metallic resources model (developed by Yukata Suzuki and his team at Osaka University).

The linkages among the three components are pictured in Figure 18.

1. *The global input-output model (GIOM)*

At the heart of the GIOM is an input-output module comprising fourteen

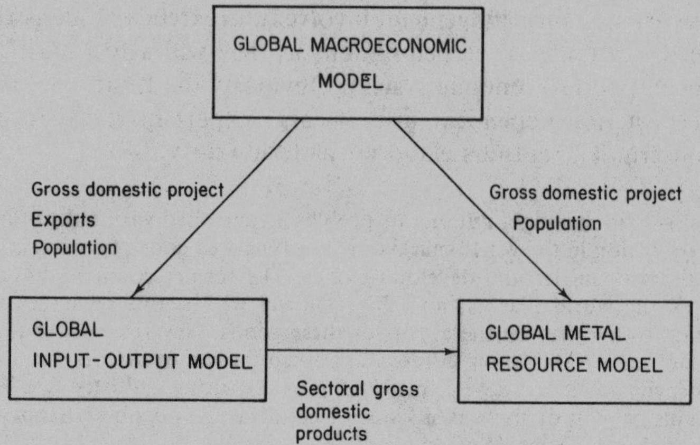

Figure 18. The basic structure of the FUGI model. (Reproduced by permission of IIASA from Kaya, Onishi, Suzuki, *et al.*, 1977)

economic sectors for each of the fifteen geographic regions in the model. This model provides, for a given year, coefficients that describe the flow of resources and finished goods among the sectors of the economy. (For example, the construction sector uses the products of the agricultural, forestry, and fishery sector and provides products to the transportation and communication sector. It may also consume some of its own products.) In addition, 'final demands', that is the sectoral outputs that are consumed or exported as end products, are calculated. Computations are facilitated by keeping track of interindustry flows in monetary, rather than physical, terms. A trade module calculates product flows among regions for each sector based on statistical (linear regression) analysis of historical data.

The final mix of sectoral value-added exports and imports which, among other things, link back into the macroeconomic model, providing a basis for long-term projections, is computed in a *linear programming module.* (Linear programming is a simplified version of the optimizing or goal-attaining approach that was described in connection with the Bariloche model.) This module is also the entry point for defining scenarios, which are objective functions that represent a possible world view or policy outcome. The model then computes, on both the macro and micro level, the detailed consequences and implications. In addition, constraints must be specified for the optimization procedure. Some of these are provided by the input-output module, trade module, and macroeconomic model, either directly or with intermediate calculations. Others are specified exogenously (by the user).

The seven constraints are:

- the balance between regional demand and supply,
- the balance between global demand and supply,

- the 'drag' involved in departing from past trends of international trade,
- the drag involved in departures from past industrial structure,
- constraints on employment in manufacturing,
- constraints on income in manufacturing,
- constraints on structural change in manufacturing industries.

2. *The global macroeconomic model (GMEM)*
The first-generation macroeconomic model* comprises the same fifteen geographic regions as the input-output model. In each region there are only six highly aggregated subsectors:

- production (measured in monetary terms),
- expenditures on gross regional product (at constant prices),
- profits and wages,
- prices (essentially a measure of price changes),
- expenditures on gross regional product at current market prices,
- official development assistance and private overseas investments (international financial flows).

The flows among the regions are trade, private overseas investment, and official development assistance. Outputs from the GMEM provide inputs that constrain the computation of input-output coefficients and, as noted above, the trade module and linear programming module of the global input-output model. Values are computed dynamically over time; that is current values depend on changes that occurred in previous years.)

3. *The global metallic resources model*
This is a system dynamics model which, like World3, views the world as a whole. The model can accept inputs from the other components of the model, but the connection is not on-line. Forecasts are made on the basis of a single, aggregated metallic resource. Factors such as available reserves, extraction costs, demand, investments, and recycling are incorporated in the structure of the model. Apparently this model does not link back to the other components of FUGI, and it is not mentioned in later papers describing the project.

Results

Like many computer models, FUGI is capable of producing a virtually infinite number of results. However, the analysts have chosen to focus on a small number of runs that explore different assumptions about relative growth in industrialized and non-industrialized parts of the world. To describe the

*A newly completed second-generation model comprises twenty-eight regions and has added a foreign aid sector to the original six.

Table 5. A comparison of 1970 gross regional products by region with forecasted values based on FUGI scenarios A and B (from Kaya, Onishi, Suzuki, *et al.*, 1977, p. 153)

		Gross Regional Product (10^9 dollars; 1970 price)							
		1970 Value		Scenario A (1985) Indexation of oil prices			Scenario B (1985) Reducing north-south gap		
No.	Region		Ratio %		Ratio %	Growth rate %		Ratio %	Growth rate %
1	Japan	189.4	(6.2)	455.5	(7.3)	(6.0)	393.0	(6.9)	(5.0)
2	Oceania	42.6	(1.4)	75.7	(1.2)	(3.9)	65.7	(1.2)	(2.9)
3	Canada	85.8	(2.8)	167.9	(2.7)	(4.6)	141.7	(2.5)	(3.4)
4	USA	977.2	(31.7)	1749.7	(27.9)	(4.0)	1513.5	(26.7)	(3.0)
5	EC	612.7	(19.9)	1136.2	(18.1)	(4.2)	926.6	(16.3)	(2.8)
6	DC O	229.9	(7.5)	459.0	(7.3)	(4.7)	373.3	(6.6)	(3.3)
7	SE.A	44.7	(1.5)	128.4	(2.0)	(7.3)	137.7	(2.4)	(7.8)
8	SW.A	78.9	(2.6)	146.8	(2.3)	(4.2)	158.7	(2.8)	(4.8)
9	M.East	23.7	(0.8)	153.3	(2.4)	(13.2)	153.3	(2.7)	(13.2)
10	Africa	27.7	(0.9)	61.0	(1.0)	(5.4)	62.4	(1.1)	(5.6)
11	L.A.	180.2	(5.9)	443.3	(7.1)	(6.2)	446.2	(7.9)	(6.2)
12	CPC.E	586.4	(19.0)	1298.6	(20.7)	(5.4)	1298.6	(22.9)	(5.4)
13	DC	2724.1	(88.5)	5342.6	(85.1)	(4.6)	4712.5	(83.1)	(3.7)
14	LDC	355.2	(11.5)	932.8	(14.9)	(6.6)	958.4	(16.9)	(6.8)
15	World	3079.3	(100.0)	6275.3	(100.0)	(4.9)	5670.8	(100.0)	(4.2)

output of the model, they differentiate between 'scenarios' and 'cases'. To define assumptions for a 'scenario', characteristics of the macroeconomic model are altered. Cases are defined by altering characteristics of the global input-output model.

The authors discuss these scenarios:

Scenario A: Indexing oil prices to the price of manufactured goods
It is assumed that oil prices and the price of manufactured goods rise at corresponding rates. The modellers regard this as a 'standard' (that is a probable and reasonable baseline) scenario.

Scenario B: Reduction of the North-South gap
In this scenario, the growth rate of advanced industrial economies is reduced by 10 per cent in comparison with *Scenario A*. This is done by shifting investments to the developing nations, where they are used to stimulate growth. The result, as one might expect, is more rapid industrialization of the developing nations.

Illustrative results for these scenarios are presented in Table 5, which compares 1970 values (in billions of US dollars, indexed to 1970) with the 1985 forecasted values for scenarios A and B.

'Cases' are defined by the authors to examine the implications for the global economy of these assumptions:

1. (Standard): essentially an extrapolation of current trends.
2. (LDC MAX): total industrial production in the developing nations is maximized.

2-1. (LDC MAX + JAPAN TRADE): same as case 2 but, in addition, Japanese exports are restricted.

3. (LDC MIN): minimization of the total production of manufacturing industries in the developing regions.
4. (PRIORITY ON COST): the costs of capital and labour are minimized throughout the world in the computations of the optimization routine.
5. (PRIORITY ON LDC EMPLOYMENT): developing countries adopt a strategy of industrial development designed to maximize employment.

Figure 19 shows one type of result that is obtainable using a combination of cases and scenarios. The motivation for this series of runs was to explore the likelihood that the developing nations could reach the goal set forth in the 1975 Lima Declaration (25 per cent of the world economy in the year 2000; 14.3 per cent in 1985). According to the analysis, only scenario B, combined with a policy of maximizing industrial development in the developing nations, offers any hope of meeting the 1985 target.

In addition to the types of results presented above, the model can also be used to forecast sectoral values added, employment in manufacturing

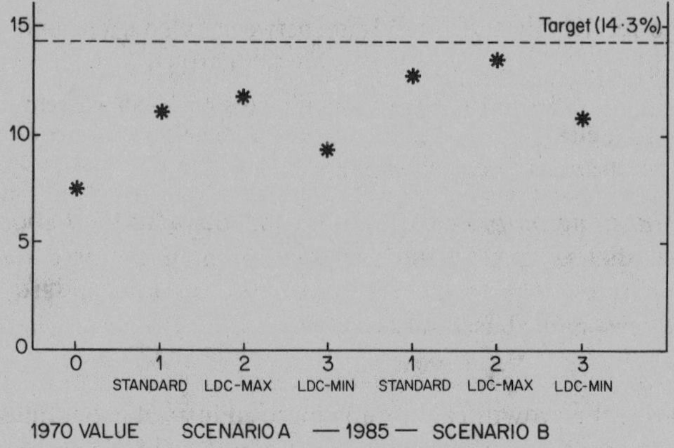

Figure 19. The share of developing countries in the world production of manufacturing industry. (Reproduced by permission of IIASA from Kaya, Onishi, Suzuki, *et al.*, 1977, p. 155)

industries, and several other issues of relevance to Japan's role in international trade.

Like several other global modelling groups, Kaya and his associates present overall conclusions that are broad and philosophical. They are based on a synthesis of the concerns that motivated the project, the model results, and what the authors learned both from the model and the experience of building it. Here is a brief excerpt:

> In the absence of a change from the traditional principle of 'survival of the fittest' to the principles of international cooperation and human solidarity, or likewise in the absence of changes from systems of wholly unrestrained free competition to systems incorporating a greater element of planning and coordination, it will probably be most difficult to overcome the various conflicts which we face in the world economy, to guarantee each country's economic welfare, and to plan for a higher degree of social welfare .... On the other hand, in a peaceful world, without such apprehensions, the possibility emerges for countries' economic security to be realized within a system of interdependence .... (Kaya, Onishi, Suzuki, *et al.*, 1977, pp. 2-12)

Here are some recommendations that are somewhat more specific:

> (1) ... we should abandon the age in which each country pursued its economic growth in its own arbitrary and self-seeking way and should be ready to meet the new age characterized by the need for international cooperation from a global viewpoint, planned coordination based on a keener feeling of human solidarity, and systems for an international control over resources.
> (2) ... we need not only to build a better system of worldwide development information to provide data on economic growth planning related to population, food, resources, energy, environment, trade and the like, but also to construct practical action plans on a global scale.

(3) ... we shall be obliged, even if it takes a great many years, to design radical new recycling systems for resource retrieval and reutilization.
(4) For the time being, there is the need to control human desires, and especially to create patterns of resource, energy, and food conservation whereby, insofar as is possible the industrially advanced countries refrain from 'overconsumption.'
(5) ... it will be necessary to build regulatory and adjustment mechanisms to insure that science and technology will be utilized for the whole of humanity ... there is a need to strengthen science-related coordinating bodies and functions in which all humankind has an interest. (Kaya, Onishi, Suzuki, *et al.*, 1977, p. 2-12)

The authors conclude:

> With the present global system, each country's relations of mutual interdependence are in fact becoming greater. In such an international environment, unless the economic policies of each individual country ... are adjusted for the world as a whole, it will probably be impossible for us to solve the problems of economic security on the global level. However, little progress will be made without a change in methods and without a reform in human value systems.

Further developments

The FUGI project still appears to be in a growth phase. Following presentation of the model at IIASA, the team was asked to develop a revision which would focus on the problems of the United Nations ESCAP (Economic and Social Commission of Asia and the Pacific) region. Voluminous projections and results were presented at the Seventh International Conference on Input-Output Techniques; in a recent personal communication, we were informed that an expansion of the macroeconomic model to sixty countries and eleven sectors is contemplated. In 1979, the Indonesian Institute of Science (LIPI) initiated a project that will use the model to study the future of the Indonesian Economy, with support from the Japan Society for the Promotion of Science. In 1980, the Australia-Japan Foundation provided a research grant to study Australian-Japanese economic relations by using the FUGI model. A project at the Free University of Amsterdam, also focusing on the ESCAP region, is collaborating closely with the FUGI team.

The authors still believe that the greatest task which global society faces is that of how to build a new international economic order. They hope that the FUGI project can contribute significantly to achieving this goal.

The principal references

Background papers

Onishi, A. (1965). 'Projections of economic growth and intra-regional trade for the developing ECAFE region 1960-1970', in *Developing Economies,* Vol. III, No. 2, pp. 158-172, IDE, Tokyo.

Onishi, A. (1971). 'A method on projections of world economic growth', *Collected Papers for Inauguration of Soka University* (in Japanese), pp. 39-47, Soka University Press.

Kaya, Y., Onishi, A., and Kono, A. (1974). North-south economic relations and industry transfer', Club of Rome Berlin Conference.

Kaya, Y., and Suzuki, Y. (1974). 'Global constraints and a new vision for development-I', *Technological Forecasting and Social Change,* **6-3.**

Kaya, Y., and Suzuki, Y. (1974). 'Global constraints and a new vision for development-II', *Technological Forecasting and Social Change,* **6-4.**

Onishi, A., Kaya, Y., and Ishitani, H. (1975). 'A. Two-level multi-nation model for development planning', in *MOIRA: Food and Agricultural Model, Proceedings of the Third IIASA Symposium on Global Modeling,* 22-25 September 1975, pp. 85-149, CP-77-1, February 1977.

Kaya, Y., *et al.* (1976). 'A step to estimating parameters of input-output tables', *Proceedings of the Fourth Global Modeling Conference,* IIASA.

Papers on global models

Kaya, Y., and Onishi, A. (1977). 'A study on the long term forecasting models for the world economy', IFAC Second International Conference on Dynamic Modelling and Control.

Kaya, Y., Onishi, A., Suzuki, Y., *et al.* in (1977). 'Report on project FUGI—future of global interdependence', Fifth Global Modeling Conference, IIASA.

Onishi, A. (1977). 'Global economic model for new international order', in *Hiroshima Peace Science,* Vol. 1, pp. 205-233, Institute for Peace Science, Hiroshima University.

Onishi, A. (1977). 'Oil price and world economy', in *Energy and World Economy in 1980s* (in Japanese), pp. 113-187, Weiss, Tokyo, March 1977.

Murakami, T., Shoji, K., and Suzuki, Y. (1978&. 'The world metallic resource model' (in Japanese), *Proceedings of Japan Society of Simulation Technology,* Vol. 6, 1, pp. 91-96, June 1978.

Onishi, A. (1978). 'A reflection on global modeling', in *The Soka Economic Studies Quarterly (in Japanese)* Vol. 7, No. 4, pp. 25-35, March 1978.

Onishi, A. (1978). 'Development of global modeling', in *Forecasts of the World Economy and Policies* (in Japanese). pp. 49-94, WEISS, Tokyo, March 1978.

Onishi, A. (1978). 'Projections of the world economy, 1977-1990', *Mita Gakkai Zasshi (Mita Journal of Economics)* (in Japanese), **71** No. 2, 48-66, April 1978.

Suzuki, Y., Shoji, K., and Murakami, T. (1978). 'Global metallic resource model', pp. 710-717, 1978 Summer Computer Conference, July 1978.

Onishi, A. (1979). *Humanistic Economics—A Guideline to the Future of the World Economy* (in Japanese), p. 238. Daisan Bunmesha, Tokyo, January 1979.

The United Nations Global Model

At the United Nations, the concept of a 'development decade' and an associated international development strategy has played a particularly important role in long-term planning. The year 1970 marked the beginning of the second development decade. Associated with this strategy and other plans and programmes put forth by various UN agencies, there have been a number of specific objectives and targets, which include these (Menshikov, 1977, p. 2):

- satisfying the basic needs of the people (in food, clothing, housing, education, medical care);

- creating one billion new jobs in the developing countries by the year 2000;
- achieving a minimal 4 per cent growth rate of food and agricultural production in the developing countries;
- increasing the share of the developing market economies in manufacturing production to 25 per cent;
- substantially reducing the average income gap between the developing and developed countries;
- improving internal income distribution so as to accelerate the eradication of mass poverty;
- achieving a new international economic order and its different components, including commodity trade stabilization, increase in financial and technological transfers, reduction of barriers to trade in manufactured goods; a code of conduct for transnational enterprise.

However, the deliberative and political process that leads to adopting these objectives provides no assurance that they will be either attainable or mutually consistent. This was a matter of serious concern to one of the agencies specifically charged with responsibility for integrated long-term planning at the United Nations, the Center for Development Planning, Projections and Policies (CDPPP).

In 1973, the Center received financial support from the Government of the Netherlands to initiate a study of the global economy, focusing particularly on environmental problems. Professor Wassily Leontief, who had received the Nobel Prize for the invention of input-output analysis and who had earlier proposed a strategy for incorporating environmental considerations into an input-output model, was named as Project Director. Subsequently, additional funding was provided by the United Nations, by the Ford Foundation, and by the National Science Foundation.

Like several others, the UN team was dispersed geographically. The major work of model building, including data collection and statistical analysis to find the best parameter values, was done by Anne P. Carter and Peter Petri at Brandeis University. The model was also programmed at Brandeis and run on the university computer. Professor Leontief, who had moved from Harvard to New York University, provided overall direction and played a role in writing the principal project report (Leontief, Carter, and Petri, 1977). Project coordination was provided by Joseph J. Stern of Harvard University. Because the study was conducted under UN auspices (although the recommendations were not to be 'official' UN recommendations), officials of the CDPPP were extensively involved in the definition of scenarios, evaluation of results, and preparation of the final report.

Peter Petri presented the UN model at IIASA in 1977, shortly after the release of *The Future of the World Economy* (the project's principal report) in New York. In presentations of the project, and especially in the popular press, strong emphasis was given to the fact that future global trends projected by this model were more optimistic than the early reports to the Club of Rome. The finding that 'no insurmountable physical barriers exist within the twentieth century to the accelerated development of the developing regions' received major attention.

Among the later models presented at IIASA, the UN world model has certainly been the most visible. Its release received front-page coverage in the *New York Times,* and magazine articles still appear occasionally which report that the model 'discredited' *The Limits to Growth.* At the United Nations, the project reports and findings have been widely circulated internationally. The model has also received considerable attention from those who are concerned with the issue that first motivated its creation, global environmental problems. In the UN General Assembly, dissemination of the project's results was followed by two important resolutions dealing, respectively, with the interrelations among population, resources, and development (RES 3345) and long-term economic trends and projections (RES 3508). These resolutions have called for additional study and research by the United Nations (Leontief, Carter, and Petri, 1977).

The purposes of the model

In describing the origins of the model we have already commented briefly on its purposes. The most comprehensive statement of purposes is found in a paper by a senior CDPPP official:

> The model was built specifically in order to provide a tool of analysis of the impact of prospective economic issues and policies on the International Development Strategy. It was to provide a framework for global projections, which would also include sufficient disaggregation and detail on the principal regions of the world; the sectoral breakdown of their economies; some basic mineral resources with special reference to energy and metals; some environmental aspects, such as pollution and pollution abatement; some detail on international trade, aid, capital flows, etc. (Menshikov, 1977).

As the model evolved, it became 'basically a general-purpose economic model', which, in Leontief's view, was 'applicable to the analysis of the evolution of the world economy from other points of view'. In a more technical article, Peter A. Petri has talked specifically about the relation between the purposes of the model and the limits-to-growth debate:

> The famous Meadows Club of Rome study was, no doubt, partly responsible for the U.N. decision to sponsor this new global model. Thus, the United Nations' original work statement spoke guardedly about the impact of environmental concerns on the goals of the second development decade. These concerns were not made explicit, but obviously referred to growing and widely-shared doubts regarding the material feasibility and social desirability of long-term economic growth.
>
> While this study has something to say about the feasibility of growth (it foresees no *physical* impediments to growth during the next twenty-five years), it

> cannot settle, and was not designed to settle the many fundamental questions raised in the *Limits-to-Growth* debate. Rather, its main contribution is to bring together, in a single manageable framework, a large body of data on economic interdependence, natural resources, agriculture, industry and the environment. (Petri, 1977, p. 261)

The structure of the model

Like the FUGI model discussed above, the UN global model is designed to use input-output analysis. The two models also have about the same number of regions (fifteen for the UN model and fourteen for FUGI). But there are important differences. The UN model's forty-five sectors for each region (including eight 'pollutants' and five pollution-abatement programmes) dwarf the fifteen that were included in early versions of FUGI (later versions include more sectors). The approach to adapting the static input-output technique to the needs of a dynamic problem is also quite different.

Figure 20 shows the structure of one region in the model. The labels in the boxes denote some of the most important variables whose values are computed by the regional model's 175 simultaneous equations. The number in the lower right corner of a box denotes the number of variables upon which computations in that box depend.

For any run of the model, fifty-four variables must be defined by the user or modeller. Some variables—for example population, labour productivity, resource reserves, and environmental standards—are always provided externally. Others—for example prices and the coefficients that govern allocating of exports—are calculated in separate submodels. In some cases, a user can decide which variables will be provided externally and which will be calculated by the model. For example, if a particular level of gross domestic product is defined as a regional goal, the model can provide information on the alternative mixes of industrial production, investment, imports, etc., that are necessary to attain this goal. On the other hand, it is possible to specify constraints such as environmental standards, import levels, and levels of development for particular industries (pollution intensive or pollution non-intensive, for example). In this case, the model projects the level of GDP and GDP per capita that is attainable in the region.

The economies of the fifteen regions are linked by flows of imports and exports, as well as capital flows. The model does not, however, keep track of flows among individual regions. Instead, demands for imported goods and exports available to meet these demands are aggregated at the global level. International prices are calculated by a separate 'satellite' model. The shares of imports and exports allocated to each region are then determined by a combination of statistical analysis and externally supplied coefficients.

Two special features distinguish this model from others that use an input-output approach. Both represent significant innovations by Leontief. The inclusion of an explicit environmental sector has already been mentioned (the UN model is one of the two global models to attempt an environmental

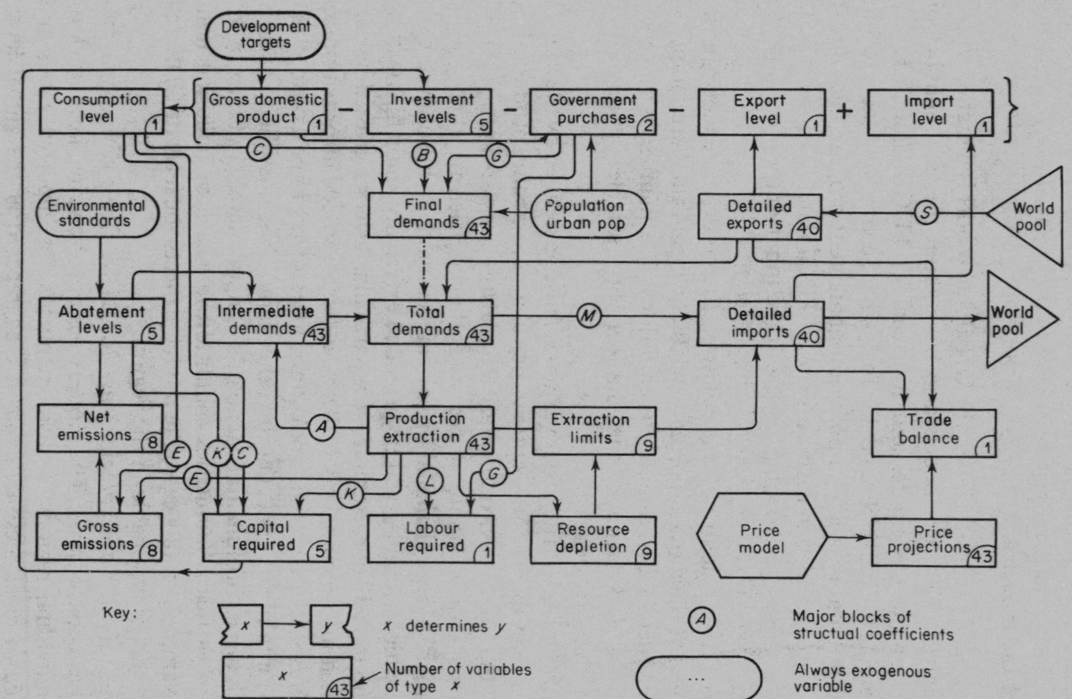

Figure 20. The internal structure of a region in the UN global model. Circled capital letters denote major structural coefficients as follows: A, interindustry inputs; B, composition of investment; C, composition of consumption; E, Generation of gross pollution; G, composition of government purchases; K, capital inputs; L, labour in industry; M, import dependence ratios; S, shares of world export markets (from Petri, 1977, p. 262)

sector). The detailed sectoral breakdown makes calculating pollutant discharges for each industrial sector in each region possible. Results are provided for emissions of particulate air pollution, biological oxygen demand, nitrogen water pollution, phosphates, suspended solids, dissolved solids, urban solid wastes, and pesticides. These are measured in physical units (millions of tons of the pollutant generated within the region in the years of the study). Abatement strategies include 'the commonly recognized treatment strategies for particulate air pollution, primary, secondary, and tertiary treatment of water pollution, and the landfilling or incineration of urban solid wastes' (Leontief, Carter, and Petri, 1977, p. 6). Particular strategies are initiated or enhanced through capital investment.

A second innovative feature of the model is its combination, within a single structure, of products measured in physical terms and products measured in monetary terms. The presence of physical commodities is especially useful in areas like agriculture, mineral extraction, and pollution abatement. Also, it provides a basis for additional crosschecking with empirical data.

The UN model is used to present a large number of scenarios for the future in terms of 'snapshots' of 1980, 1990, and 2000. However, it differs from every other model discussed in that it cannot, in any sense, be viewed as a dynamic model. Scenarios are generated based on statistically estimated parameter values for 1970 and externally supplied values for the appropriate future years. This characteristic of the model is elaborated by Menshikov (1980):

> ...at present, a solution for any given year is practically independent from the solutions for any previous year, with the exceptions of the base year (1970). A small change from 'historic' (1970) capital in the previous period (t-T) would be sufficient to make the model partially dynamic in some of the scenarios.
>
> The structure of the model dictates the type of scenarios for future projections. It is obviously not fruitful to use the model for short- or medium-term growth paths, since it is not based on behavioral equations estimated from time series. Nor is it possible, at present, to use the model to forecast trends based on an optimization procedure with a set of interchangeable objective functions.
>
> However, because of its inherent stability and fairly detailed structure, the model may be useful for analytical studies of possible future long-term trends, for change in technology and consumption patterns, resource availability, change in international relationships, etc. For each set of assumptions, the model will provide a global outcome, which can be analyzed from the point of view of feasibility and consistency. (Menshikov, 1980)

Results

Among the global models described here, the popularly oriented report on the UN model devotes the most attention to presenting detailed results. Most of them are based on a relatively optimistic scenario (scenario X), which is one of four that are described in detail. The scenarios are defined by specifying aggregate growth rates of population and GDP per capita for the 'developed countries' and the 'developing countries' as given in Table 6.

Table 6. The growth rates in population and GDP per capita for developed and developing countries for four scenarios used in connection with the UN global model (from Leontief, Carter and Petri, 1977, pp. 30-31)

	Growth rate: population		Growth rate: GDP per capita	
	Developed	Developing	Developed	Developing
Scenario A (endogenously computed growth rates)	0.7	2.3	3.2	3.1
IDS scenario: minimum international development strategy targets for developing countries; extrapolation of long-term growth rates for developing countries	1.0	2.5	3.5	3.5
Scenario X	0.7	2.3	3.3	4.9
Scenario C	0.6	2.0	3.0	4.9

Scenario A can be viewed as an 'old economic order' scenario. In it there is no provision for any substantial increases in internal or external investment rates (which would be required to accelerate growth in the developing areas); also, apart from extrapolating current trends, no provision is made for increases in export shares and import substitution in the developing countries, which would make possible alleviating their balance-of-payments problems (Leontief, Carter and Petri, 1977, p. 30).

The IDS scenario is based, as noted, on the minimum-growth targets in the developing countries. However, this turns out to be a relatively pessimistic scenario. Because of the differential growth rates in population, the expansion of gross product per capita is no faster in the developing nations than in the world as a whole. Consequently there is no change in the 12 to 1 ratio of per capita GDP between developed and developing nations during the period 1970-2000.

The X and C scenarios were designed by the modellers to explore the implications of more optimistic assumptions for the international and regional economies. 'In these basic scenarios, growth rates of gross product *per capita* are set in such a way as to roughly halve the income gap between the developing countries by 2000 with a view towards closing the income gap completely by the middle of the next century. 'It is further assumed that the growth rates of per capita gross product decline in the developed countries when higher absolute levels of income are achieved. The respective growth rates are based on the 'medium' UN projection for scenario X and the 'low' UN projection for scenario C.

The results discussed in *The Future of the World Economy* include:

- assessments of regional differences,
- internal structural changes in regional economies,
- food and agriculture,
- mineral resources,
- changes in the structure of international trade,
- international service, aid, and capital flows,
- balance-of-payments problems.

Because it is an area covered by other models, we will present, for illustrative purposes, detailed results for food and agriculture. Then we will summarize some of the more important general conclusions that have emerged from work with the UN model.

Table 7 gives data on agricultural output for scenario X, Table 8 shows data

Table 7. Agricultural output for scenario X of the UN global model (from Leontief, Carter, and Petri, 1977, p. 39)

	1970	1980	1990	2000
Total agricultural output (billions of 1970 dollars)				
Developed	255.8	296.0	332.2	415.4
Developing group I[a]	23.8	42.1	78.5	151.7
Developing group II	120.1	186.4	322.8	529.9
World	399.7	524.5	733.5	1097.1
Regional share in world output (percentage)				
Developed	64.0	56.4	45.3	37.9
Developing group I	6.0	8.0	10.7	13.8
Developing group II	30.0	35.5	44.0	48.3
World	100.0	100.0	100.0	100.0
Agricultural output per capita (billions of 1970 dollars)				
Developed	0.2615	0.2776	0.2878	0.3381
Developing group I	0.0665	0.0884	0.1222	0.1770
Developing group II	0.0526	0.0652	0.0910	0.1227
World	0.1104	0.1192	0.1372	0.1713
	1970-1980	**1980-1990**	**1990-2000**	**1970-2000**
Rate of growth of agricultural output[b] (percentage)				
Developed	1.4	1.2	2.2	1.6
Developing group I	5.7	6.2	6.6	6.4
Developing group II	4.3	5.4	5.0	5.1
World	2.7	3.4	4.0	3.4

[a]Developing group I is countries with major mineral resource endowments; development group II is the other developing countries.
[b]Average annual compound.

Table 8. Land requirements and land productivity in 2000 for scenario X of the UN global model (from Leontief, Carter, and Petri, 1977, p. 40)

Region	Agricultural output	Land/yield index	Arable land	Land productivity
Developed market				
North America	196	215	111	194
Western Europe (high income)	130	162	100	162
Japan	176	269	100	269
Oceania	192	296	183	162
Centrally planned				
Soviet Union	164	215	100	215
Eastern Europe	143	186	100	186
Asia (centrally planned)	488	333	120	278
Developing market				
Latin America (medium income)	495	517	166	311
Latin America (low income)	532	460	140	328
Middle East	950	612	126	487
Asia (low income)	506	376	113	331
Africa (arid)	409	371	131	282
Africa (tropical)	438	492	152	324

on land requirements and land productivity in the year 2000 for scenario X, and Table 9 shows a comparison for 1980 of various agricultural factors for scenario X and one (labelled G) that assumes a self-sufficiency in food in low-income Asia. These tables only begin to show the richness and diversity of the results obtainable from an input-output model. For example, scenario X projects for the year 2000 roughly a fivefold average increase in agricultural output in the developing market, led by a nearly tenfold increase in the Middle East; increases in productivity and land under cultivation both contribute to this increase. The comparison of scenarios in 1980 shows that only a relatively modest increase in productivity would be required to achieve self-sufficiency in food.

We conclude our discussion of the conclusions derived from the UN model with a selection of the findings presented in the excellent summary in *The Future of the World Economy:*

> 1. The target rates of growth of gross product in the developing regions set by the International Development for the Second United Nations Development Decade are not sufficient to start closing the income gap between the developing and the developed countries.
> 2. The principal limits to sustained economic growth and accelerated development are political, social, and institutional in character. No insurmountable physical barriers exist within the twentieth century to the accelerated development of the developing regions.

Table 9. Comparisons of factors related to agricultural production in 1980 for scenarios X and G of the UN global model (from Leontief, Carter, and Petri, 1977, p. 41)

Factors related to agricultural production	Scenario X	Scenario G	G − X	Percentage change $\frac{G-X}{X} \times 100$
Agricultural output (millions of dollars)	60.6	62.6	+2.0	+3.3
Land/yield index (percentage)	141.0	147.7	+6.7	+4.8
Land investment (millions of hectares)	1.1	2.3	+1.2	+109.1
Irrigation investment (millions of hectares)	0.4	0.8	+0.4	+100.0
Imports of fertilizer (millions of tons)	1.93	1.99	+0.06	+3.1
Pesticide use (millions of tons)	0.39	0.40	+0.01	+2.6
Capital stock agriculture (millions of dollars)	5.4	5.6	+0.2	+3.7
Net food imports (millions of dollars)	1.7	0.0	−1.7	−100.0
Trade balance (deficit) (millions of dollars)	−10.2	−7.9	+2.3	−22.6

[a]Scenario G assumes hypothetical self-sufficiency in food in low-income Asia.

3. The most pressing problem of feeding the rapidly increasing population of the developing regions can be solved by bringing under cultivation large areas of currently unexploited arable land and by doubling and trebling land productivity. Both tasks are technically feasible, but are contingent on drastic measures of public policy favorable to such development and on social and institutional changes in the developing countries.
4. The problem of the supply of mineral resources for accelerated development is not a problem of absolute scarcity in the present century but, at worst, a problem of exploiting less productive and more costly deposits of minerals and of intensive exploration of new deposits...
5. With the current commercially available abatement technology, pollution is not an unmanageable problem...
6. Accelerated development in developing regions is possible only under the condition that from 30 to 35 per-cent and in some cases up to 40 per-cent of their gross product is used for capital investment. A steady increase in the investment ratio to these levels may necessitate drastic measures of economic policy in the field of taxation and credit, increasing the role of public investment and the public sector in production and infrastructure. Measures leading to a more equitable income distribution are needed to increase the effectiveness of such policies; significant social and institutional changes will have to accompany these policies...
7. There are two ways out of the balance-of-payments dilemma [posed by potentially large deficits in the developing regions]. One is to reduce the rates of development in accordance with the balance-of-payments constraint. Another way is to close the potential payments gap by introducing changes into the economic relations between developing and developed countries, as perceived by

the Declaration on the Establishment of a new International Economic Order— namely, by stabilizing commodity markets, stimulating exports of manufactures from the developing countries, increasing financial transfers, and so on.

8. To accelerate development, two general conditions are necessary: first, far-reaching internal changes of a social, political, and institutional character in the developing countries, and, second, significant changes in the world economic order. Accelerated development leading to a substantial reduction of the income gap between the developing and developed countries can only be achieved through a combination of both these conditions. Clearly, each of them taken separately is insufficient, but, when developed hand in hand, they will be able to produce the desired outcome. (Leontief, Carter, and Petri, 1977, pp. 10-11)

Further developments

The team that developed the UN global model is no longer functioning as a unit. However, the model lives on in three incarnations.

One version remains at Brandeis University. Peter Petri and Anne Carter are using it to address theoretical and technical issues involved in global input-output modelling. It is interesting to note that they have opted to simplify their version, rather than make it more complex. They have dropped the environmental sector because of data problems.

A second version is still available at the United Nations and is used occasionally for analysis by CDPPP.

A third version, located at New York University, is being expanded, refined, and applied under Professor Leontief's direction. Most recently two studies of the worldwide economic implications of military spending have been completed for the UN and US disarmament organizations (see Duchin, 1980). A study on the future production and consumption of non-fuel minerals is in progress and nearing completion.

The principal references

Leontief, W., Carter, A., and Petri, P. (1977). *The Future of the World Economy,* a United Nations Study, Oxford University Press, New York, 1977. (This book has been published in translation in many languages by various publishers.)

Carter, A., and Petri, P. (1975). 'Resources, environment and the balance of payments: applications of a model of the world economy', Delivered at the Third Reisenbury Symposium on the Stability of Contemporary Economic Systems, July 1975. (Publication forthcoming in a conference volume.)

Petri, P. (1976). 'Some economic implications of pollution-intensive exports' in *Environmental Modeling and Simulation* (Ed. W.R. Ott), pp. 282-287, United States Environmental Protection Agency, Washington, D.C.

Carter, A. (1977). 'Oil prices, terms of trade and regional development prospects in the world economy', Statement before the Subcommittee on Energy of the Joint Economic Committee, Congress of the United States, 13 January 1977, Government Printing Office, 1977.

'We've called you here today to announce that, according to our computer, by the year 2000 everything is going to be peachy.' (From *The New Yorker*)

Menshikov, Stanislav M. (1977). 'Using the global input-output model for long-term projections', Prepared for delivery at the Fifth IIASA Conference on Global Modeling, UNC DPPP.

Petri, P. (1977). 'An introduction to the structure and application of the United Nations world model', *Applied Mathematical Modeling,* **1**, No. 5, June 1977, 261-267.

Carter, A., and Petri, P. (1978). 'Factors affecting the long term prospects of developing regions', Economic Council of Canada discussion paper, Ottawa, Canada.

Duchin, Faye (1980). 'The world model: inter-regional input-output model of the world economy', Presented at the Eighth IIASA Conference on Global Modeling, July 1980.

Menshikov, Stanislav M. (1980). 'Using the global input-output model for long-term projections', in *Input-Output Approaches in Global Modeling* (Ed. G. Bruckmann), Pergamon, New York.

Petri, P. (1981). 'An algebraic framework for interregional modeling', *Journal of Regional Science* (forthcoming).

How Global Models Differ and Why

This section compares the seven models in order to emphasize similarities and differences. While some of these have already been noted— or may have occurred to you in spite of our failure to mention them— we think that a conscientious listing here will be helpful in preparation for Chapters 4 and 5. Chapter 5, in fact, will discuss the issues raised here in much greater detail.

The basic character of a model— and its points of difference with other models— may be largely shaped even before the project begins. The answers to the five questions that follow set the context for a model development project and impose constraints upon it. Sometimes these questions are forgotten in later stages of the project.

1. Who conceived the idea of a model and who wanted it to be built?
2. What specific problems, if any, was it supposed to address?
3. What preconceptions and experiences did the modellers bring to the project?
4. Who paid for the work to be done and what did they think they were paying for?
5. Who was supposed to use the model after it was completed and for what purpose?

After those crucial questions have been answered, the next questions are usually resolved by the modelling team. Since no two projects are exactly alike, it is hard to generalize about how this is done. Sometimes the decisions about these questions made during the modelling process are instinctive and unrecognized. Often team members are hard put to reconstruct exactly what happened after the project is over.

1. Should we use a model? (Actually, modellers almost never ask this question.)
2. What paradigms will we rely on? Will we use the same one for each sector?
3. How many sectors should we have and what should they be?
4. What things should be computed within the model, what should we include by defining variables externally, and what should we leave out entirely?
5. How many regions should there be and how should we define them?
6. How long a period of time should each 'cycle' of the model represent?
7. How far into the future should we allow the model to run?
8. To what degree should we rely on formal mathematical analysis to solve' the model? To what degree should we rely on simulation?

There are some important technical differences among the models that we will not summarize here. Some of these have already been discussed— and they will be taken up again by the modellers and evaluators (and by us) in the following chapters. They include:

1. The degree of reliance on 'hard' data.
2. The extent and type of documentation.
3. The use of sensitivity analysis to test the model's behaviour.
4. The extent and type of validation.
5. The specific conclusions and recommendations.
6. The form of presentation and use of the model, if any.

Who paid for what and why?

Many observers have argued that a policy-oriented model will be most useful when

- it focuses on a specific, well-defined problem;
- this problem reflects the objectives and needs of a specific client;
- the sponsor (provider of funds) and the client (intended user) are the same.

Table 10 classifies the seven models according to these criteria. The table shows that none of the models met all three criteria at the time of their presentation at IIASA (the FUGI model does now). The Mesaroviç/Pestel model, which has attracted a great deal of attention from policy makers at high levels, is intended to be a general-purpose model, and therefore has neither a specific client nor a well-defined problem (Mesarovic views this as a strength of the model). The three models that appear to have the best survival potential (SARU, FUGI, and MOIRA) either began with government support

Table 10. The seven models classified according to the three criteria of usefulness

The model	Focused on a fairly specific and well-defined problem	Oriented towards the objectives and needs of a specific client	The client and the source of funding are the same
Forrester/Meadows	×	×	
Mesarovic/Pestel			
Bariloche	×		
MOIRA	×		×[a]
SARU		×	×
FUGI	×	×[b]	×[b]
UN world model		×	×
Number of cases	4	4	4

[a]Since 1976 work on the MOIRA model has been supported by the Netherlands ministries of Agriculture and Development Cooperation.
[b]The FUGI model is now being partially supported by the UN Economic and Social Commission for Asia and the Pacific and applied to the ESCAP region.

or have it now. With the exception of SARU and the UN global model, every project has had problems either raising initial funding, securing continuing funding, or both. Most funding sources have not been willing to support the sort of speculative venture that global modelling seems to be. Funders (as well as most research institutes) feel more secure supporting work on problems that they believe *can* be solved than on problems that *ought* to be solved.

The academic disciplines of the core team members

Table 11 summarizes information about the disciplines of the core team members.

Table 11. The academic disciplines of the core team members of the seven modelling groups. The figures in the table give the numbers of team members in each category

	World modelling team							
Academic discipline	F/M	M/P	Bar	MO	SA	FU	UN	Total
Engineering, computer science	1	3			2	5		11
Management science/systems development	2							2
Management science/operations research					1			1
Economics			1	4	1	2	3	11
Biological sciences	1	1	2	5	1			10
Mathematics			1					1
Physics	2			1	1			3
Social sciences[a]		2	1					3

[a]Including political science, sociology, psychology, and education.

Engineering and economics are the disciplines most strongly represented, while there has been essentially no participation by other social scientists (Richardson and Hughes, who worked on the Mesarovic/Pestel project, were both trained as political scientists, but did not function in this capacity as members of the project team). One may conclude from these data that the global models are mechanistic and simple when it comes to matters of politics, culture, social organization, and the like. Indeed, most observers, including participants in the Sixth IIASA Symposium on Global Modeling, would agree. In most of the models (even those developed by engineers), economic variables have been given overwhelming emphasis.

On the other hand, in most of the projects there was an awareness that social, political, and cultural variables are important— indeed, inclusion of these variables was usually discussed. However, such reasons as the following were advanced for excluding them:

- There were no data.
- Social scientists who were queried did not have any good ideas about how these 'soft' factors could be incorporated in the model.
- The modellers felt that putting these factors into the model would lessen its credibility.

Influential modelling paradigms

We have already said a great deal about modelling paradigms, and, although the topic interests us, we will not say much more here (we will return to this subject in Chapter 6). Table 12 summarizes the key information.

Table 12. The modelling paradigms that influenced the development of the seven world models (× indicates a primary influence, an × in parentheses indicates a secondary influence)

	The world model							Totals	
The paradigm	F/M	M/P	Bar	MO	SA	FU	UN	Primary	Secondary
System dynamics	×	(×)			(×)	(×)		1	3
Economics/econometrics[a]		(×)	(×)	×	×	(×)		2	3
Optimization			×	(×)				1	1
Input-output analysis[a]		(×)				×[b]	×	2	1
Eclectic	×[b]							1	0

[a]Both input-output analysis and economic/econometric analysis fall within the perview of the discipline of economics.

[b]Both the FUGI and M/P models are highly eclectic; however, in the case of the FUGI model, input-output analysis is clearly the dominant paradigm. The M/P model is difficult to classify; however, economic/econometric approaches are becoming increasingly dominant in later versions.

If you had heard about global modelling only from the—mostly hostile—reviews in the economic literature, this table would be surprising, for it shows that the field is dominated by traditional approaches based on economics. Although many persons assume that global models and system dynamics are inextricably linked, the table also shows that this is a misperception; indeed, system dynamics has been less influential than its proponents had hoped and its detractors had feared. This latter point becomes even clearer if it is recognized that the inclusion of the Mesarovic/Pestel and SARU models under the system dynamics label would almost certainly be disputed by their project leaders. Nevertheless, we believe that these projects were influenced by system dynamics, even though the models are not system dynamics models. The FUGI project's use of system dynamics is limited to one sector, energy, which is not well integrated with the rest of the model.

The consequences of economists' growing involvement in global modelling is partly reflected in the tables that follow. More recent models tend to have more narrowly defined boundaries, shorter time horizons, and more disaggregation. Economists tend to select variables for inclusion in their models on the basis of theory and convention within their discipline, rather than relevance to the problem at hand. When data are missing, economic modellers are more likely to exclude relevant variables than to supply the necessary values from their own—or anyone else's—intuition, experience, or judgement. As the influence of economists in the field of global modelling has grown, concern with broad global issues derived from the problématique of the Club of Rome has diminished.

Note to the reader

After we had written the last paragraph, we realized how clearly it reflected our biases. Accordingly we decided that it would be fair to supply an alternative, as follows. You may choose whichever you prefer.

As a result of the growing involvement of economists in the field of global modelling, there is a healthy trend to a more solid theoretical base and a stronger commitment to scientific method in the development of models. Recently the models have tended to have boundaries that are more narrowly defined, and to have shorter time horizons and more disaggregation. A variable is only included when there is a sound theoretical justification for doing so and when appropriate data to support parameter estimation are available. The zealotry and leaps of faith that characterized the early global models have now been supplanted by more prudence, more rigour in analysis, and sounder scientific practice.

What gets put in and what gets left out: model boundaries

Modellers often use the concept of a boundary to distinguish the variables that are computed within the model (endogenous variables) from those that are read in from external sources (exogenous variables), or those that are ignored

Table 13. A rough comparison of the seven global models on the basis of their treatment of nine key variable classes as endogenous, exogenous, or omitted. The following code is used in the table:

× denotes that at least one variable in this class is computed endogenously
X denotes that exogenous variables in this class are defined
○ denotes that this sector is omitted from the model

Variable class or sector	F/M	M/P	Bar	MO	SA	FU	UN	Totals ×/X/○
Population	×	×	×	X	X	X	X	3/4/0
Energy and mineral resources	×	×	○	○	×	×	×	5/0/2
Agricultural sector								
Consumption/demand	○	○	×	×	×	×	X	4/1/2
Production	×	×	×	×	×	×	×	7/0/0
Non-agricultural sector								
Consumption/demand	○	○	×	○	×	×	X	3/1/3
Production	×	×	×	○	×	×	×	5/1/1
Prices/market mechanisms	○	×	○	×	×	×	X	4/1/2
Trade	○	×	×	×	×	×	×	6/0/1
Environmental pollution	×	○	○	○	○	○	×	2/0/5

entirely. The choice of what to make endogenous or exogenous or to ignore is one of the most crucial decisions in model making; it has a major effect on what sorts of conclusions it will be possible to draw.

Table 13 makes a rough comparison of the seven global models on the basis of endogenous, exogenous, and omitted variables. Column entries in this table identify, for each model, sectors with at least one internally computed variable. By examining the rows, it is possible to determine which sectors are most frequently incorporated in the models and which have been given relatively less emphasis. We hasten to emphasize that this picture is hardly complete. A particular entry can refer to a complex input-output table or a single equation. Also, we have not mentioned possible sectors such as, for example, domestic violence, probability of war, disarmament, or political stability, which are not included in any of the seven global models.

Geographical and sectoral levels of aggregation

There is no easier way to criticize a model than to suggest the desirability of greater disaggregation or of including an omitted variable or sector.

However, every modelling effort is subject to severe constraints of time and resources. It is never possible (or desirable) to include everything that the analysts— or others— think may be desirable. As a model becomes more detailed, data requirements, mathematical problems, and overall complexity increase at a rapid rate. Detail can become a heavy burden when the time comes to debug the computer program or write documentation. Sometimes the

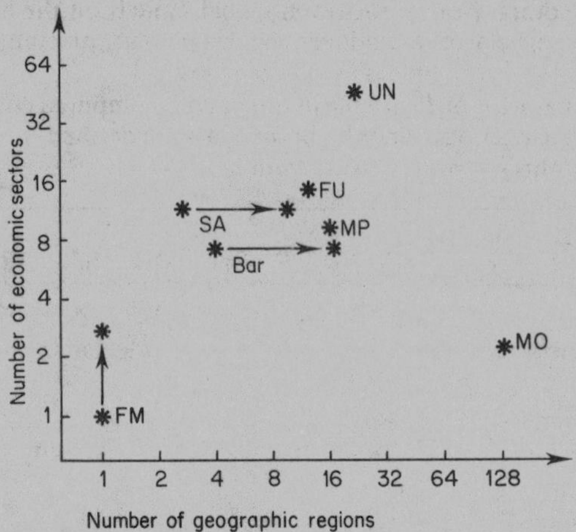

Figure 21. The levels of disaggregation of the seven global models with respect to geographical regions and economic sectors. The arrows in the figure denote increasing numbers in later versions of the models

decision to use a high level of disaggregation results in excluding whole sectors or, perhaps, limiting the time available for validation, sensitivity analysis, documentation, or even actual use of the model to calculate results. It can also make running the model very expensive.

Figure 21 shows some data about the levels of disaggregation of the seven global models with respect to economic sectors and geographical regions.

Analytic 'solutions'

With the exception of World3 and the Mesarovic/Pestel model, all of the global models use analytic procedures to some extent. The Bariloche model, which uses mathematical programming techniques to generate 'optimal' policies, is the most strongly oriented in this direction. The MOIRA model, with its highly detailed international trade mechanism, is the most elegantly complex from a mathematical standpoint.

In general, the more analytically-oriented global models tend to be more complex, more difficult for policy makers to understand, and less inclusive, although this need not be the case. Also, there is a greater commitment to research and theory development in the later models. The proper use of analytical techniques in the field probably reflects a more mature stage of development than the exclusive use of simulation or scenarios; the premature use of such techniques may, however, create a false sense of certainty about some results.

Decisions about levels of aggregation and using analytical approaches dictate the amount of computer time needed for a single run to a significant extent. Thus, these factors can affect the way the model is used and by whom on technical grounds as well as intrinsic biases. The highly aggregated World3 model is the simplest and easiest to run; the Mesarovic/Pestel and SARU models are also fairly close to this end of the spectrum of complexity, while the more complex MOIRA and Bariloche models take much longer to run.

The simpler models permit running a large number of scenarios and a wide variety of sensitivity analyses (although such explorations of variation are, alas, not as often done as they should be, even with simple models). With these models a single 'solution' or result is regarded as less important than the overall feel for a problem area that can be gained from a large number of interactions with the model. With large econometric and optimization models, relatively fewer runs are likely to be made, with greater importance attached to each run.

Time horizon

By a model's time horizon we mean the last year in the future to which the model can be run. Among the seven models, World3 has the longest time horizon (to 2100) and FUGI has the shortest (to 1985). The time horizon may be an important factor in distinguishing between optimistic and pessimistic futures. Both World3 and the Mesarovic/Pestel models reveal problems that may occur some time after the turn of the century. Models that have shorter time horizons cannot shed much light on such problems. On the other hand, the more analytically oriented and detailed econometric and input-output models are probably more useful for short- and mid-term projections.

A concluding comment on diversity

Reading over this chapter, you may have been struck by the diversity of these seven models, by their very different problem definitions, methods, degrees of aggregation, and ideological casts. Or you may have seen not diversity but similarity—the common use of material and economic variables, the dependence on a single UN data base, the general engineering/management approach, and the surprisingly uniform conclusions, given the different geographical and ideological starting points. Both characteristics, diversity and similarity, are there to be seen. Perhaps one of the greatest lessons to be learned as the number of global models increases is how human beings from various parts of the globe see the world differently and how they see it commonly.

How many worlds are there?

How could a supposedly scientific activity like computer modelling
addressing what is presumably the same globe
produce such different descriptions
of how things are
and how they might be?

Here are three possible answers to this question.

Take your pick, or
think up a better one.

ONE: THE MODELS ARE DIFFERENT BECAUSE THEY HAVE DIFFERENT PURPOSES.

A scientist wondering about servomechanisms
watches a frog tongue tracking a fly.

A second scientist, interested in differentiation,
sees an egg turn into a tadpole and then a frog.

A third scientist, concerned about nerve function,
traces electrical signals through the nerves and muscles of a frog's leg.

. . . Same frog.

Global models describe the world in different ways because those who build them have different purposes.

TWO: GLOBAL MODELLING IS NOT A SCIENCE. IT IS MORE LIKE A CRAFT.

What global modellers do is more like

pottery		astronomy
architecture		physics
cabinetmaking	than	chemistry
weaving	like	biology
and bonsai		or mathematics

A model is a synthesis of science, art and technology.
The modeller is the synthesizer.

Like any craftsperson, a good modeller must:

have talent
have a sense of esthetics
have a sense of function
be intelligent
be creative
be humble
practice a lot
work hard
and express his or her individuality.

Every potter makes teapots
but not all teapots are alike.
You wouldn't want them to be.

THREE: NOT EVEN SCIENTISTS SEE THE SAME WORLD OBJECTIVELY.

We all actively create the worlds we see.

We decide how the world is, and lo!

we see the part of it that is that way,
we don't see the part of it that isn't that way,
we befriend the people who see it our way,
we avoid the people who see it another way.

Global models are wonderful expressions
of how some people in the world
see the world.

The global model you have decided you like best may tell you
not so much about the world
as about yourself.

4

The actors have their say: the questionnaire responses by the global modellers

Before the Sixth IIASA Symposium on Global Modeling convened, the global modellers were sent a questionnaire. It was answered by representatives of the seven modelling groups. This chapter presents the questionnaire and the responses.

The method of responding differed from group to group, as did the degree to which the responses represent the group as a whole. In some cases, the responses were derived from an extensive discussion; in others, the task of responding was entrusted to a single author. Since modelling teams are not necessarily more homogeneous than other human groups, it may well be the case that some participants in the projects described do not agree fully with everything that their colleagues wrote in response to our queries. Thus, we have indicated to the best of our abilities before each response who supplied it and to what extent it was an individual or a group effort.

The Questionnaire

Introductory remarks

All the following questions relate to *global modelling*. For the purposes of this enquiry, the specific features of a global model (as distinct from a regional or sectoral model) are

1. its unique kind of 'closedness' (many fewer exogenous variables than in regional or sectoral models),
2. its comprehensiveness (economic *and* metaeconomic variables), and
3. its time horizon (> thirty years).

The questions are deliberately formulated in a future-oriented (rather than in a past-oriented)way. They are grouped under seven headings, which they serve to illustrate without necessarily being complete. *We ask you to reply on the basis of your own experiences gained with your model, but not limited to this experience alone.*

1. The purposes and goals of global modelling

What are the main problems (what is the single most important problem) a global model should try to analyse? To what extent can global modelling serve this purpose?

What are the specific features of goal-setting in a global modelling effort? How should normative aspects interrelate with descriptive aspects? What should be the predictive value of a global model? Under what conditions? If no absolute prediction is possible, how should the model be put to use?

What kind of 'global goal function' could be conceived or what other possibilities of representation of goal-seeking behaviour do you see in global modelling?

How would you proceed if you had (a) limited resources or (b) (practically) unlimited resources to spend on global modelling work?

What services might (will) global modelling be able to render in the future? What services not?

2. Methodology

Given the purposes spelled out under part 1, what are the consquences with respect to methodology? For model structure? For mode of model use? How far do certain methodologies reflect/determine certain world views?

In particular:

- How should aspects be handled that one knows to be important but about which one lacks data or knowledge of relations (for example environment or 'the human factor')?

- Many modellers have rather uncritically used cross-sectional, static data to estimate what are essentially longitudinal, dynamic relationships. Do you perceive this as a major problem? Are there strategies for getting around it or must we live with the constraint of insufficient time-series data?
- How can non-material needs be represented?
- How can physical constraints be best treated (limiting values, for example the proportion of carbon dioxide in the atmosphere, the rate of heat dissipation, etc.)?
- How should the price system be taken care of?

3. Actors, policy variables

Who are (a) the actors in the model whose behaviour is endogenously modelled, (b) the actors steering the model, and how are they represented?

How does the model handle policy variables? Does the model assume that policies remain constant unless specific policy intervention is made during the course of the run? Or does the model automatically change policy variables during the course of a run, thus giving the appearance that problems have vanished without any specific action on the part of national leaders? What policy options can the model test?

How do the model policy options relate to the policy options and choices being made by leaders today? How many of the policy variables correspond to policy alternatives that political decision makers can actually select, given their power, and how many are of a more abstract, synthetic nature?

How should political constraints of an institutional kind be tackled methodologically?

What do the structural equations depict in terms of goals of actors?

How do the actors interact?

4. Structural aspects

Regarding past modelling efforts, could one have arrived at the same results (i) with a smaller model or (ii) with a different model (approach)?

Have you developed a formalism by which to decide what to leave out of a model?

Have you made any effort, after the model was constructed, to compress and simplify its essence in a form that would allow for understanding rather than simply complex explanation? What insights does the model yield that would not have been available through other means of analysis?

How is consistency among regions guaranteed?

How does the model represent the fact that actors adapt to changes?

How should changes in structural relationships be considered? How should technological progress/technological change be represented? What other types of structural relationships are important in this context? Could 'structure sensitivity analysis' help to answer these questions?

Should a model be set up as a one-time venture or should it deliberately be designed in a modular fashion, flexible enough to adapt to new insights and/or new policy questions? How can such flexibility be ensured?

5. Testing the model

Which aspects of your model do you have much confidence in and which aspects do you have less confidence in? Which aspects of your model require the most subjective judgement? Which aspects are the most concrete? What kind of errors are to be expected (size and frequency)? What assumptions in your model are you least certain about?

What provision have you made to take the uncertainty of the parameters into consideration and the uncertainty in the data and in the structural equations? Has your model produced sensible results when subjected to noise and larger disturbances?

How can one be assured that the model will correspond to reality (validation)?

How can one be assured that the model did correspond to reality (calibration)?

What is the role and purpose of sensitivity tests?

6. Internal organization

How should global modelling work by organized?

How can consistency between subgroups be secured?

How should global modelling groups cooperate in the future?

What rules and procedures would you lay down to ensure that documentation of the model is kept continuously up to date?

What are the best ways of interweaving with the scientific community at large?

7. Relations between modeller and user

Should an explicit client be identified at an early stage?

To what extent did clients participate in the work? What formats to communicate with the clients should be developed?

What can be done to help the client to avoid misunderstandings and to understand the strengths, weaknesses and limitations of the model?

Which clients need world modelling most? Which are most interested in world modelling?

Do you have recommendations on how to reduce 'overselling'?

The Forrester/Meadows Models

Source of the response

The response dealing with the Forrester/Meadows models was prepared by Donella H. Meadows, Associate Professor of Policy Studies and Environmental Studies, Dartmouth College, Hanover, New Hampshire, USA. Dana Meadows was a member of the Core group of the World3 project, with particular responsibility for the aspects of the model dealing with population and nutrition. She was also the principal author of *The Limits to Growth* and, of course, she is a co-editor of this book. Her contribution reflects her own view, as well as that of Dennis Meadows, but was not reviewed by Jay Forrester or other members of the MIT modelling team.

Because readers have been widely exposed to the controversy surrounding *The Limits to Growth*, they have undoubtedly also formed some conclusions about the creators of the World3 model. Therefore, many will be surprised to discover that Dana is one of the most skeptical of the global modellers.

1. The purposes and goals of global modelling

a. What sorts of questions are accessible to global (that is closed, interdisciplinary, long-term) modelling approaches and what are not? What sorts of predictions can be made with global models?

Karl Deutsch divides the future of any system into two distinct regions. In the *short term* the system's behaviour is dominated by the momentum of past decisions and actions. Natural delays in the system, levels of stocks and inventories, materials in pipelines, buildings under construction, contractual obligations, the information in people's heads— all affect the system and none can be changed quickly. During the short term the system can hardly be affected, but it *can be predicted,* at least theoretically.

In the *long term* the system becomes more and more free from its historical constraints, and a very wide range of outcomes is possible. It is still subject to immutable physical, geological, and social laws, but it depends much more on the policies imposed between now and then. Because of this freedom, absolute prediction is impossible, but policy intervention (beginning well in advance, before the long term becomes the short term) can have major effects. That is, in the long term, the system *can be affected,* and because it can be affected, it cannot be predicted.

The dividing line between the short and the long term depends on the natural delays in the system. For the human metabolic system the short term may last from a few hours to a few months. For the global socioeconomic system the short term probably does not extend beyond five years, and the long term begins perhaps after twenty years.

Global models deal with a long-delay system, and presumably attempt to assess the impact of deliberate policy on that system. Therefore global models should not be expected to deliver absolute, precise predictions. At best global models can provide conditional, imprecise predictions; if a given set of policies is imposed, the general tendency of the system will be such-and-so.

Even with the limited range of conditional, imprecise predictions, a global model *cannot* provide answers to all questions, or even to very many questions at one time. The essence of modelling is simplification that can lead to understanding. A model that is designed to answer too many questions will be so complex that it is as incomprehensible as the real system and thus must be treated as a 'black box'. Global models, more than any other kind, suffer from the expectations of both modellers and policy makers that all the world should be contained within the model— that the model should answer a global range of questions rather than a single question of global scope. The best global models will be those that are carefully constructed to answer a sharply defined, limited set of questions. That means that there should be very many global models.

Global models, like all formal models, may also provide some services that are not immediately related to policy questions. They can serve as clear, unambiguous representations of points of view, as crosscultural, crossdisciplinary communications devices. They can help identify inconsistencies and gaps in knowledge of a given system. They can be basic research tools, allowing the expression of a hypothesis in a falsifiable form, the testing of its logical consequences, and subsequent modification of that hypothesis.

I personally believe that human understanding of human socioeconomic systems is still so faulty that the basic research function of computer modelling is by far the most important.

b. What are the most important problems accessible to the field that have not yet been analysed by global models?

A very long list of questions could be included here. Very few of the important holistic questions that could be asked about the global system have ever begun to be answered. Here are a few that especially bother me:

- What are the real causes of social maldistribution of important commodities such as food, energy, and capital?
- What is the actual economic effect of world armaments production? What is the most likely political outcome? How could one begin to affect the system so that it will generate a *dis*armament race?

- How can the world system move in smooth progression from nearly total dependence on petroleum to nearly total dependence on renewable energy sources such as solar, wind, and biomass?
- How resilient are global ecological cycles to man-made impositions? What early warning signals might give timely information that further stress could lead to irreversible change?
- Could an economic system be designed that meets all basic human needs and that does not depend on perpetual physical growth?
- What is the impact of changing global communication patterns on social evolution?
- What might be the effects of massive defaults of loans to Third World countries on the international monetary system?
- Is the influence of multinational corporations on world trade likely to increase or decrease? What will be the consequences of either outcome?
- How do ideas, concepts, and world views affect the workings of human socioeconomic systems? Are there useful policy levers in the sphere of human ideas?

c. Is there a role for normative models on the global level? Can a global goal function' or objective function for an optimizing model be conceived?

There is indeed a role for normative modelling on any level, if normative modelling is taken to mean the expression of goals and visions of desirable futures. As Elise Boulding has said, 'imaging the future' or 'conceptualizing the totally other' is a necessary skill for any society that expects to evolve, to solve its long-term problems, to create, and to progress. Formal mathematical modelling could enhance this skill by providing social visions that are more explicit, comprehensive, and internally consistent than verbally expressed visions can be.

This does not mean, however, that formal optimization models with 'global objective functions' have an obvious role in global modelling. I believe that optimization methods are invaluable for sorting out detailed, short-term policy choices where the system boundary is quite limited, and the goals, choices, and constraints are commensurate, clearly understood, and uncontroversial. I do not believe that these conditions apply to most global problems.

Applying optimization methods to questions such as the ones posed in the previous section is likely to produce an apparently precisely chosen path where precise knowledge is in fact unavailable, to confuse real policy levers with mathematically convenient but fictional ones, and to oversimplify the multitudinous, interacting, conflicting goals of human societies. Optimization is an extremely powerful tool for fine-tuning policy decisions. For most global questions, where there is still argument about the basic structure of the system and the best general direction to aim for, optimization is not yet an applicable technique.

d. Can any aspect of goal-seeking behaviour be incorporated in a global model?

Yes, the *actual* goal-seeking behaviours of individuals and institutions within any system can and should be included in any model that tries to explain the behaviour of the system or to simulate how the system will respond to new conditions or policies. The basic premise of systems theory is that human beings are goal seeking and that they respond to whatever information is available about the system's state in order to adjust that state nearer to their own desires. The challenges in modelling a social system are to discover what information actors are really responding to, what goals they are actually pursuing, and what really happens when different actors responding to different information and pursuing different goals come into conflict. I don't see how any social system can be modelled usefully without addressing these questions. I also believe more can be learned by incorporating the theories of goal seeking explicitly in the model rather than leaving them to be introduced by man-machine interactions.

e. How would you proceed if you had (i) limited resources and (ii) practically unlimited resources to spend on global modelling work?

If I had limited modelling resources, I would begin by defining the model's question extremely carefully. I would then make the very simplest model possible that encompasses *all* the interacting factors I felt were important to the question. This first model would be very aggregated and crude, but it would contain at least some rough representation of each relation I felt was important to the system. Whatever resources remained would then be devoted to disaggregating, adding detail, and estimating parameters more precisely until either the model became satisfactorily realistic or (more likely) half the resources were exhausted. The other half would be reserved for documenting the model.

If I had unlimited modelling resources, I would seek mightily to limit them, since I believe the above procedure is desirable in any case. I would allocate excess resources to different models addressing different questions or, even better, to different modelling groups from various cultures and ideologies, using different methods, but all addressing the same question. I would then give each model to another group for testing and evaluation.

2. Methodology

a. Do methodologies reflect/determine certain world views?

Yes, each different modelling school is derived from a different perception of the world and limits or constrains its practitioners' approaches to problems. The constraining influence of a modelling method is unavoidable. Human beings are unable to think without some such conceptual constraints. All problem-solving efforts, whether mental or formally mathematical, are limited by the set of a priori expectations about the world that has been labelled world view, paradigm, or gestalt.

There is no way to transcend this limitation, but its effect can be minimized by a careful matching of the real-world problem with whatever modelling technique is most closely suited to the characteristics of the system that generates the problem. This matching process requires a broad knowledge of many disciplines and modelling techniques, deep introspection on the part of every modeller concerning his/her own biases, methodological as well as ideological, and an extremely careful definition of the problem to be modelled. I would say that the single worst problem in the field of modelling at present is the inexperience of modellers in examining their own assumptions, realizing their lack of objectivity, and understanding the relative strengths and weaknesses of their methods.

b. What should be done with elements of a system known to be important but difficult to model because of lack of data or hypotheses?

My modelling school (system dynamics) is biased towards including such elements in the model, even if their nature or mathematical forms must be based on intuitive judgement. There are several reasons for this bias:

- Omitting such elements implies that they are unimportant, which is often the one thing about them we know is not true. Therefore even an approximation will yield better results than total exclusion.
- In models of the sort we make (non-linear, causal, strongly closed-loop simulation models), inclusion of a structurally correct feedback loop, even with somewhat inaccurate parameters, is often sufficient to give us the results we need (reproduction of dynamic tendencies rather than precise prediction). This reason is *not* necessarily applicable to other methods that seek other kinds of results.
- The purpose of a model is not to dictate the truth but to put forth a hypothesis for discussion and for attempts at disproof. Therefore if one believes an unmeasured psychological, environmental, or political factor is important in a system, one should state this belief explicitly and precisely, to encourage observation or experimentation that will increase knowledge about this factor.

- Inclusion of assumptions about poorly understood relationships will permit the sensitivity testing required to determine whether it is useful to acquire improved knowledge about the relationship.

c. Are there problems with using cross-sectional versus longitudinal data for parameter estimation, and are there ways to avoid these problems?

This question is not of much interest in the system dynamics method, because it is subsumed by a much larger question — what is the role of formal statistical estimation procedures in modelling? We do not make much use of statistical techniques, because they are not easily applied to non-linear feedback systems and because we do not often need the fine degree of accuracy given by a least-squares fit rather than a visually approximated one. We do use both cross-sectional and longitudinal data, in addition to any other source of information we can find, including anecdotal verbal descriptions and direct observation of decision makers within the system. If there is any estimation rule within our method, it is simply to use all information skeptically, intelligently, and carefully — to be aware of how it was obtained, how it may be biased, and what messages can be drawn from it with certainty and what cannot.

As a practical matter, the number of important parameters in a global model for which both time-series and cross-sectional data are available over a period comparable to that of the projection is very small. In any event, global models are typically used for projecting the system into modes of behaviour so different from any on which we have good data that the validity of regression estimates is very low — so low that it obscures the technical differences between the two data types.

d. How can non-material human needs be represented?

This is a vitally important question. If the modelling and policy community is unable to answer it, the likely result will be a society that meets material needs and no others. I believe that analysts have a great responsibility to include in their models some indices of 'truth and beauty' outputs of the social system as a continuous reminder that such factors are of major concern. For example, if I were to judge the success of any policy or social order, I would like some indication, however crude and imprecise, of such factors as:

- quality of the environment (purity and beauty),
- quality of employment (dignity and challenge),
- equity (fairness and justice),
- freedom of choice (opportunity, diversity, and tolerance),
- sustainability (security, resilience to external perturbations),
- fulfillment (self-actualization, spirituality),
- knowledge (understanding, absence of self-deception),

as well as material sufficiency. These are all terribly difficult factors even to discuss, much less measure. But every person is able to discriminate at least the extremes of their presence or absence; disagreement comes only when fine distinctions or difficult choices are required. All analysts know *something* about non-material needs, and they know they are important. Our reluctance to incorporate them in our models may reflect our desire to appear omniscient, objective, and scientific, which we falsely correlate with numerical precision.

e. How can physical constraints be best treated?

To omit environmental constraints is to select for policies that pretend there are no physical limits — the one thing we are sure is not true, especially in the long term. Perhaps the best way to incorporate uncertain physical limits is to test the model with high, low, and intermediate estimates to see how much difference they might make. These high and low estimates are subject to much bias and should probably be obtained from people or groups with clear optimistic or pessimistic casts (for example the Hudson Institute versus Paul Ehrlich). It is also important to represent physical constraints as dynamic, not static. Aside from the rate of input of solar energy, all global physical constraints I can think of are in dynamic interaction with human activities; physical limits can be raised or lowered by human intervention.

f. How should the price system be taken care of?

Any question that is the subject of as much continued controversy within the modelling community as this cannot have an easy or final answer. The representation of the market within any social system model must depend entirely on the question, time horizon, and geographical scope of the model. I have made relatively short-term (five to twenty-year) models where goods and money markets, inventories, and pricing systems had to be represented in excruciating detail, because the dynamics of the system and the policies to be tested demanded that detail. In our global model the price system is represented extremely simply because the question we were asking required representation of only its gross dynamic effects:

> World3 includes several causal relationships between the supply of some material (such as food, nonrenewable resources, or industrial capital) and the response of the economic system to that material's scarcity (develop more agricultural land, allocate more capital to resource production, or increase manufacturing efficiency). These relationships are most realistically represented with price as an intermediate variable:
> decrease in supply → rise in price → social response.
>
> World3 simplifies the model of social response by eliminating explicit reference to price, the intermediate variable. We shortened the representation of the causal chain to:
> decrease in supply → social response.

> The ultimate regulating effect of the price system is thus included, but price does not explicitly appear in the model.
>
> We assumed that the price system conveys its signals of scarcity to decision makers accurately and with a delay that is insignificant on a 200-year time scale. Thus the price mechanism was eclipsed to increase the model's simplicity and comprehensibility. If actual prices are biased or do not immediately reflect declining resource supplies in the real world, the global economic system will be less stable than its facsimile in World3. If price information is transmitted to institutions that adjust their production or consumption patterns only after a long delay, the economic system will be less able to adjust itself to any limit, and the tendency for the system to overshoot its limit will be increased. Our representation of the price system in World3 thus tends to overestimate the stability of the real-world system. (D.L. Meadows *et al., Dynamics of Growth in a Finite World,* p. 17, MIT Press Cambridge, Mass. 1974).

I suppose that, if any generalization can be made about market representation in social system models, it would be that one should avoid the tendency to choose representations that are mathematically convenient rather than appropriate to the task of the model. The frequency of free-market formulations in current computer models is simply not in accord with the scarcity of actual free markets on this planet. The free-market representation is only appropriate as a very simple approximation when market dynamics are irrelevant to the problem being examined. Otherwise, unfortunately for the ease of the modelling task, much more emphasis has to be placed on who actually sets prices, what information they respond to, how long the response takes, what constrains the range of prices that could be set, and how long it takes for consumers to perceive and respond to a new price.

g. Why did you choose the method you used in formulating your global model?

As in virtually all other modelling efforts, the method came first and shaped the question we asked, rather than vice versa. We were practising system dynamicists, attuned to observing long-term basic dynamic trends in social systems, to asking qualitative, conditional, and imprecise policy questions, and to emphasizing the parts of the global socioeconomic system that are dominated by non-linear, delayed, feedback relationships. The purpose of our model (see Chapter 3) was exactly matched to this method, of course. If we had not been system dynamicists, we would not have asked such a question.

3. Actors, policy variables

a. Who are the actors in your model whose behaviour is determined endogenously?

In system dynamics, endogenous policies and decisions are embodied in *rate equations,* which are often complex compilations of the aggregate results of decisions made by many different actors. In World3 some major rate functions, the actors implied by them and the major elements affecting them are (this list is illustrative, not inclusive):

Rate	*Actors*	*Major endogenous inputs to rate*
Birth rate	All families	Population age structure, fertility control technology, general level of health, level of industrial development
Death rate	General population, health care agencies, governments	Food per capita, health services per capita, pollution
Industrial and service capital investment rate	Investors, governments, consumers	Level of industrial development, availability of food, services, resources, and manufactured products relative to desires
Non-renewable resource usage rate	Producers, consumers	Population, industrial output per capita
Pollution generation rate	Producers, government	Population, per capita resource usage rate, agricultural production
Land development rate	Farmers, government	Potentially arable land, marginal productivity of land development
Land erosion rate	Farmers, government	Intensity of farming

In every case the behaviour is modelled as adaptive to the state of the system and as directed to some goals.

The major policy assumed endogenously throughout the model is that industrial growth, economic expansion, and the 'westernization' of values will be encouraged to proceed along roughly historical lines throughout the globe until interrupted by some unavoidable physical obstacle. The model system automatically adjusts itself, allocates outputs, and uses resources according to simple economic guidelines in pursuit of industrial development as long as it can. It also assumes that families make reproductive decisions based on their own economic needs rather than on the state of the whole system. These assumptions can be overridden by exogenous policy changes, as discussed below.

b. Who are the actors steering the model? How do the model policy options relate to the choices being made by leaders today? How many of the policy variables correspond to policy alternatives that policy makers can actually select?

Examples of *exogenous* decisions or policies that are formally incorporated in World3 and the actors implied are:

Policy	*Possible actors*
Reduced rate of natural resource use per unit of industrial output	Governments enforcing recycling, technical change agents, consumers choosing conservation-oriented products
Reduced rate of pollution generation per unit of industrial output	Governments enforcing pollution standards, technical change agents, consumers insisting on 'ecological' products
Increased land yields per unit of agricultural input	Plant breeders and agronomists, pest controllers, foundations, and governments investing in 'green revolution' programmes
Improved fertility control	Governments or other institutions promoting family planning or investing in human fertility research
Decreased desired family size	Educators, propagandists, cultural change agents
Stabilization of capital investment at a desired output level	Governments, consumers stabilizing their own material desires

Note that these are large, aggregate policies and the details of how they are enacted are not specified. The model is intended to ask *whether* a recycling, pollution-control, or family-planning programme would have a desirable effect, not to determine how such a programme could be carried out.

c. Does the model assume that policies remain constant unless specific policy intervention is made, or does it automatically change policy variables during the course of the run?

Either, at the discretion of the model user. We have tested most of the above policies, and others, in two forms: imposed exogenously with no cost and no delays, and incorporated endogenously to turn on at an exogenously determined state of need (when resource costs exceed a certain level, for instance) and with capital costs and development delays. There is a striking difference in the results using the two forms (see Chapter 7 of *Dynamics of Growth in a Finite World*). The outcomes are much more favourable using the purely exogenous formulations of the type reported in *The Limits to Growth*.

d. How should institutional and political constraints be tackled?

This question is similar to the one about the price system, and the answer must be similar. There is no one best way; each modeller must choose the way most appropriate to the purpose of his model.

However, two common ways of representing these constraints can be identified as probably *in*appropriate, especially for long-term global models. One is to ignore political constraints completely. Like physical constraints or intangible social/psychological factors, the one thing we know about these factors is that they are often important, and the one formulation that is most likely to be wrong is the assumption that they are not present. The second serious mistake is to assume that political constraints are unchangeable and will always be as they are today. Some models are even based on the assumption that, in a conflict between physical constraints and political ones, the physical laws will change to accommodate political desires, rather than vice versa.

In any case, questions of what kinds of pressure will cause human institutions to change, and how and at what rate they may change, seem of vital importance to the evolution of the global socioeconomic system. Hypotheses about institutional change should be brought to the forefront of long-term social system modelling instead of being hidden in the background.

e. What do the structural equations depict in terms of goals of actors in your model?

In the difficult task of modelling human values we tried to include only the most basic values that can be regarded as globally ubiquitous. They begin with requirements for survival — food, water, shelter — and go on to include a hierarchy of other desires — longevity, children, material goods, and social services such as education. Some of these values (for example, desired family size and preferences among food, material goods, and services) are represented explicitly in the model as variables that have an important influence on economic decisions. Others are included implicitly as in the allocation of service output to health services or in the quantity of nonrenewable resources used per capita.

All the values included in World3 are assumed to be responsive to the actual physical and economic condition of the system: thus they are all involved in feedback loops. The patterns of dynamic value change included in the model, however, were limited to the patterns of change historically observed over the last hundred years or so. During that time, the major global force behind value change has been the process of industrialization, a process that is still under way in most nations of the world. Therefore, the values that both shape and respond to the development of the model system follow the historical patterns observed during industrialization. As industrialization increases in the model — measured by the level of industrial output per capita — the aggregate social demand shifts in emphasis from food to material goods and finally to services. Changes also occur in the emphasis placed on obtaining children, education, and health care and in the distribution of various goods and services throughout the industrializing economy.

We did not build into World3 any global shifts in values other than those that might be expected to take place as the world becomes more industrialized. It is possible that new value systems will evolve, but the pace of value change is always slow and the precise nature of new social attitudes is currently more a matter for speculation than for logical projection on the basis of any generally accepted body of theory. We chose to represent unprecedented value changes by a set of switches throughout the equations of the model, which the operator can use to express his own hypotheses about future social development. (*Dynamics of Growth in a Finite World,* p. 19)

f. How do the actors interact?

In any feedback model, actors, by their decisions and actions, affect various system states, and the conditions of these system states are then used as information that guides further decisions and actions. Typical system states in World3 are population, capital, cultivated land, pollution, and natural resource reserves. Different actors throughout the system respond to these various system states and try to adjust them to meet various goals. These goal-directed actions are not always successful, and sometimes are in conflict.

Industrial output, for example, can be used to attain goals in the agriculture, service, resource, and capital sectors. If there is not sufficient output to meet all goals, some must be postponed, or some expectations must be revised. In the case of allocation of output, the conflict is resolved in World3 by an assumed hierarchy of goals. The resource sector needs are filled first (otherwise all production could stop) and then consumption needs (an implicit assumption about the political power of consumers). The remainder is divided between food and service needs, depending on the relative distance of each sector from its (variable) goal. If there is a dire food shortage, for example, most of the output goes to agriculture. The residual after agriculture and service investments have taken place goes to further investment in the industrial sector.

The total pattern of interactions in World3 is quite complex, even though the model is aggregated and conceptually fairly simple. The result of the various actors and sectors each acting to influence some subset of system states in response to information about some other subset of system states is a total behaviour that is usually not intended or desired by any particular actor in the system.

4. Structural aspects

a. Could one have arrived at the same results with a smaller model or a different modelling approach?

Jay Forrester's World2 model leads to very similar conclusions and is considerably smaller than World3 (see J. Forrester, *World Dynamics,* Wright-Allen Press, Cambridge, Mass., 1971). *After* making World3 we learned enough about its essential mechanisms to make a very condensed model that exhibits the same basic behaviour. We would not have been able to produce such a condensation without constructing the larger model first, of course, nor would we have had much confidence in such a crude model. Nor is it very useful for elucidating different kinds of policies.

Many other people have arrived at the basic conclusions of World3 via mental models, although their expression of these conclusions was necessarily less precise and systematic (see our own conclusions in section *d* below). A very similar set of conclusions was also reached by Mesarovic and Pestel using simulation but not system dynamics. Models using other techniques have not produced the same results, but they have been asking different questions that are better suited to their different modelling techniques. I believe that many modelling approaches are philosophically incapable of producing, or at least unlikely to produce, results similar to those of World3.

b. How do you decide what to leave out of a model?

System dynamicists take this question very seriously, since our methodological bias favours simple, transparent models. Our primary techniques for keeping models simple are probably training and peer pressure. We are just as likely to attack each other for including unnecessary complications as for omitting some vital factor, and a sure way to win praise within the field is to find some elegant way of representing simply a system that was previously represented in a cumbersome or disaggregated fashion. Students are forced to limit their initial models to only one or two levels or feedback loops, and the goal of transparency is reiterated constantly.

Our formal methods for attaining simplicity include:

- Careful and deliberate definition of the problem, including a sketch of the expected behaviour of the main variables over time (reference mode). Inclusion or omission of a variable is then always related to whether it could possibly influence the reference mode.
- Initial production of a crude but comprehensive running model that approximates the reference mode. We aim to have this model running within a month or so of the beginning of any project. Ideally this first model will contain the minimum causal structure necessary to produce reasonable behaviour. It is then elaborated, disaggregated, and expanded until we begin to have confidence in its structure. Every expansion must be proven necessary, either because it enhances the dynamic capabilities of the model or because it increases the model's credibility to the client. We dislike adding dynamically unnecessary

elements just to please clients, but it is often easier to do so than to argue against it.

- For difficult cases we have to use trial-and-error procedures. For example, in the World3 model I was unsure whether a population age disaggregation would be important to the overshoot dynamic of the model. I constructed four different population structures, disaggregated into 1, 4, 15, and 65 different age categories. After testing them all, I concluded that the four category disaggregation was sufficient for our purposes (see *Dynamics of Growth in a Finite World,* Chapter 2).

c. *Have you made any effort after the model was constructed to simplify its essence in a form that would allow for understanding rather than simply complex explanation?*

Yes, that was exactly the purpose of *The Limits to Growth.*

d. *What insights does the model yield that would not have been available through other means of analysis?*

The most general insights to be drawn from World3 are these:

> The dominant behavior mode of World3 is caused by three basic assumptions about the population-capital system:
>
> 1. The prevailing social value system strongly favors the growth of population and capital. Therefore, these quantities tend to grow unless severely pressed by physical limitations. Their growth is exponential because of the inherent positive feedback nature of industrial production and human reproduction.
> 2. Feedback signals about the negative consequences of growth are generated by the environmental systems that support population and capital. These signals take the form of pressures against growth, such as diminishing returns to investment in agricultural inputs, the buildup of harmful pollutants, increased development costs for new land, increased resource costs, and less food per capita. The negative feedback signals become stronger as population and capital grow toward environmental limits.
> 3. Delays in the negative feedback signals arise for two reasons. First, some delays, such as those inherent in population aging, pollution transfers, and land fertility regeneration, are inescapable consequences of physical or biological laws. Second, some delays are caused by the time intervals necessary for society to perceive new environmental situations and to adjust its values, institutions, and technologies in response.
>
> A system that possesses these three characteristics — rapid growth, environmental limits, and feedback delays — is inherently unstable. Because the rapid growth persists while the feedback signals that oppose it are delayed, the physical system can temporarily expand well beyond its ultimately sustainable limits. During this period of overshoot, the short-term efforts required to maintain the excess population and capital are especially likely to erode or deplete the resource base. The environmental carrying capacity may be so diminished that

it can support only a much smaller population and lower material standard of living than would have been possible before the overshoot. The result is an uncontrollable decline to lower levels of population and capital.

With this understanding of the system characteristics that lead to instability, it becomes relatively easy to evaluate alternative policies for increasing stability and bringing about a sustainable equilibrium. For example:

1. Short-term technologies design to mask the initial signals of impending limits and to promote further growth will not be effective in the long term. Rather, they will disguise the need for social value change, lengthen the system's response delays, and increase the probability, the speed, and the magnitude of the eventual overshoot and collapse.
2. Policies that combat the erosion of the earth's resource base will certainly reduce the severity of decline after an overshoot. However, so long as growth is still emphasized and feedback delays persist, resource conservation will not in itself prevent overshoot. Furthermore, the overstressed system may not be able to afford the costs of conservation during a period of overshoot.
3. Social value changes that reduce the forces causing growth, institutional innovations that raise the rate of technological or social adaptation, and long-term forecasting methods that shorten feedback delays may be very effective in reducing system instability.
4. A judicious combination of policies designed to prevent the erosion of resources, foresee the effects of approaching limits, and bring a deliberate end to material and demographic growth can circumvent the overshoot mode altogether and lead to a sustainable equilibrium.

Although these conclusions seem to be simple and self-evident, most economic and political decisions made today are based implicitly on a world view very different from the one presented here. The dominant contemporary model contains the assumptions that physical growth can and should continue; that technology and the price system can eliminate scarcities with little delay; that the resource base can be expanded but never reduced; that the solution of short-term problems will yield desirable long-term results; and that population and capital, if they must ever stabilize, will do so automatically and at an optimal level. (*Dynamics of Growth in a Finite World,* pp. 561-652)

e. How is consistency among regions guaranteed?

Regions are not disaggregated in World3.

f. How are changes in structural relationships considered?

In some cases model relations are added or subtracted by means of exogenous policy switches. An entire feedback-regulated pollution-control sector, for example, can be activated or not at the desire of the model operator.

The entire model structure evolves continuously through time because of the prevalence of non-linear relations. The relation between food and mortality can shift, for instance, from being strongly dominant in the system when food is very scarce, to being one of several equivalent mortality determinants, to being ineffective when food is in abundance. Similarly, the various physical limits in the model may be completely ineffective or totally dominant, and their dominance can shift from one limit to another, depending on the state of other factors in the system.

g. *How should technological change be represented?*

This is another question whose answer depends upon the purpose of the model. This is how we answered it:

> Technological advance, like price, is a social phenomenon; it results from applying man's general knowledge about the world to the solution of specific, perceived human problems. If we were to make a complete dynamic model of the development of a given technology, we would include the following:
>
> 1. A level of accumulating general knowledge, with the rate of accumulation dependent on the resources devoted to basic research.
> 2. A widespread perception of some human problem.
> 3. An allocation of physical resources, human effort, and time to search for a technical solution to the problem. If the level of basic knowledge is great enough, the solution will be found after some delay.
> 4. Another delay, to allow for testing, social acceptance, and implementation of the new technology. The length of this delay will depend on the magnitude of the required departure from the present way of doing things.
> 5. A larger impact of the technology on the total system, including social, energy, and environmental costs in areas widely separated from the actual point of implementation.
>
> Nearly every causal relationship in World3 could conceivably be changed by some sort of new technology. In the past, various technologies have directly or indirectly improved birth-control effectiveness, increased food production, and provided better techniques for abating pollutants. They have also increased life expectancy through medical advances, hastened the rate of land erosion, and developed more toxic pollutants. It is by no means certain that technologies will continue to do any of these things in the future, for the human values and economic institutions that govern technological development are always subject to change.
>
> Since so many causal relationships can be altered by conceivable technological changes, we had to consider building technological change directly into each model relationship as we formulated it. We did that by assigning possible technologies to three categories: already feasible and institutionalized, feasible but not institutionalized, and not yet feasible.
>
> Some causal relationships have historically been influenced by technology and continue to be so influenced today. They occur whenever social agreement is reached about the desirability of change and when resources and institutions to bring about that change are already integral parts of the system. Some examples are medical technology to improve health, industrial technology to raise production efficiency, agricultural technology to increase land yields, birth-control technology to control family size, and mining technology to discover and exploit lower-grade nonrenewable resources. A significant number of the world's people have adopted a value system that will continue to promote these technologies as long as their costs can be afforded. They are effectively built into the world's socioeconomic system. Therefore, they are also built into the relationships of World3, with the assumption that they will continue to develop and spread through the world, without delay, as long as economic support for them continues.

Other technologies have not been so widely accepted that they can be considered functioning parts of the world system. It is not clear that all the nations of the world will be willing to institutionalize and pay for such technologies as pollution control, resource recycling, the capture of solar energy, supersonic transports, fast-breeder reactors, or increased durability of manufactured goods. All these technologies are feasible, but it is not possible to state with confidence when or whether any of them will be adopted on a worldwide scale. Therefore, we incorporated many of them as optional functions that anyone analyzing World3 can activate at any specified time in the model run. The model can thus be used to test the possible impact of any or all of these technologies and the relative advantages or disadvantages of adopting them.

One set of technologies was omitted from the model — discoveries that we cannot possibly envision from our perspective in time. No model, mental or formal, can incorporate unimaginable future technologies as they may actually occur. That is one reason why no model can accurately predict the future. Any long-term model that is being used to aid the policy-making process must be updated constantly to incorporate surprising new discoveries as they are made and to assess how they may change the options of human society. (*Dynamics of Growth in a Finite World,* pp. 17-19)

h. Should a model be set up as a one-time venture or should it deliberately be designed in a modular fashion, flexible enough to adapt to new insights or new policy questions?

So much time and effort is involved in making a credible model that one is tremendously tempted to keep on using it once it is completed, adding to it and adapting it as new information or problems come to us. Some kinds of models for some purposes can indeed be used in this way — the large programming models used to control complex industrial production routines or shipping routes, for example. But these models are, in fact, confined to systems of limited scope and relatively great certainty, and the decisions derived from them are generally short term. The relations in them are straightforward and uncontroversial, so one can have faith in the general correctness of the model even if it is very complex.

For large-scale, long-term social problems, I believe the modular, general-purpose sort of model is a mistake. It quickly becomes a monster, enormously complex and incomprehensible even to its makers. It loses two of the main advantages of formal modelling — transparency relative to the complicated and baffling real system, and openness to criticism. Because of the state of understanding of social systems, its relations are bound to be speculative and subject to much discussion. Believing the results of such a model is an act of faith that no wise policy maker is willing to commit himself to.

Thus, there is a conflict between the role of a model as an explicit hypothesis to be comprehended, tested, and perhaps discarded, versus the understandable desire of modellers to see their labours result in widely useful products with long lifetimes. I would suggest that one way to reduce the conflict is to regard the output of modelling to be general ideas and concepts, rather than precise prescriptions for action. With the more general goal of improving the mental models of decision makers, much smaller models could be made. Each model could be focused on a clear problem definition, rather than dealing with all problems. The labour per model would decrease, and then the willingness of modellers to see their hypotheses disproved and to come up with new and better hypotheses would increase.

5. Testing the model

a. *Which aspects of the model do you have most and least confidence in?*

We have great confidence in the basic qualitative assumptions and conclusions about the instability of the current global socioeconomic system and the general kinds of changes that will and will not lead to stability (see the summary in section 3). We have relatively great confidence in the feedback-loop structure of the model, with some exceptions which I list below. We have a mixed degree of confidence in the numerical parameters of the model; some are well-known physical or biological constants that are unlikely to change, some are statistically derived social indices quite likely to change, and some are pure guesses that are perhaps only of the right order of magnitude. The structural assumptions in World3 that I consider most dubious and also sensitive enough to be of concern are:

- the constant capital-output ratio (which assumes no diminishing returns to capital),
- the residual nature of the investment function,
- the generally ineffective labour contribution to output.

b. *Which aspects of your model are most subjective?*

Probably the most subjective part of any model is the choice of the problem and method. Also, in any long-term global model the parameters expressing earthly limits and the effects of technologies upon such limits must necessarily be fairly uninformed estimates — no one really knows what the carrying capacity of the planet might be or even which factors are most likely to be limiting. We dealt with the representation of physical limits as follows:

> The negative feedbacks that can balance the growth potential of population and capital are contained primarily in the agriculture, resource, and pollution sectors of the model in the form of assumptions about the physical limits of the global system. The limits are represented as dynamic, not static. They may be raised or lowered, depending on events elsewhere in the system. World3 incorporates the following assumptions about these limits:
>
> 1. The amount of potentially arable land that can be developed into actually cultivated land through the investment of capital is finite. As the stock of potentially arable land is diminished, the marginal cost of land development, measured in terms of capital and energy, increases.

2. There is a limit to the amount of food that can be produced from each hectare of arable land each year. This limit can be approached by investment in agriculture inputs such as fertilizers, pesticides, and tractors. Eventually, however, there are diminishing returns to these inputs. The land yield limit can be decreased by very high levels of pollution or overintensive cultivation, and it can be restored to its original value by investment in land maintenance.
3. The stock of nonrenewable resources in the earth is finite. The absolute limit of available resources is the entire mineral content of the earth's crust. However, long before that limit is reached, the marginal cost — in capital and energy — of extracting and processing each unit of resource will rise to prohibitive levels.
4. There is a limit to the rate at which environmental pollutants can be rendered harmless by natural assimilation processes. Pollution levels can be kept below that limit by reducing the toxicity or quantity of industrial and agricultural emissions. However, pollution control requires capital investment, which is subject to diminishing returns. Moreover, the rate of natural pollution absorption can be lowered by the pollutants themselves if they reach levels that interfere with the environment's assimilation mechanisms...

For the reference run of the World3 model, we attempted to assign values to the limit parameters that are consistent with present global resource estimates and currently foreseeable technologies. Other values, both higher and lower, were tested in subsequent model runs. The values assumed in the reference run are:

1. Potentially arable land— 3.2 billion hectares, or about twice the area currently under cultivation.
2. Maximum yield per hectare— 6,000 vegetable-equivalent kilograms per hectare-year, or three times the global average yield in 1970.
3. Nonrenewable resources (total exploitable stock)— 250 times the amount consumed globally in 1970.
4. Persistent pollution assimilation rate (per year)— 25 times the amount of pollution rendered harmless by natural ecosystems in 1970.

We believe that the assigned values represent an intermediate position between the extreme ecological and technological views. They allow for considerable progress beyond the limits attainable with present technology, but they do not assume that technology will be able to push back the physical limits indefinitely. (*Dynamics of Growth in a Finite World,* pp. 15-16, 23)

The acrimonious debate surrounding our model indicates that carrying-capacity estimates must be extremely subjective. I can only explain the intensity of that debate by assuming that willingness to accept low or high limits is closely connected with one's culture, ideology, and even personal identity. Fortunately, the numerical estimates about physical limits in World3 can vary over a rather wide range (± at least one order of magnitude) without changing the general, qualitative conclusions of the model.

c. What kind of errors are to be expected?

This question is difficult to answer in a standard numerical, statistical form, since we make no claim to any precise predictive ability at all. Our model question concerns only the qualitative behavioural characteristics of the system — under what conditions is it likely to be stable or unstable? We believe we have captured the basic behavioural tendencies of the system correctly, with the exception of those parts of the unstable runs that enter a collapse mode. Our knowledge is almost certainly insufficient to represent the system's response to a complete reversal of historic trends.

d. What provision have you made to take the uncertainty of the parameters and structural equations into account?

We have done very extensive sensitivity testing, much of which is reported in *Dynamics of Growth in a Finite World.* Each individual sector has been tested, as well as the model as a whole. Structural relations and parameters have been changed singly and in groups. The model has been forced into extreme conditions, feedback loops have been strengthened or entirely broken, incomes have been held constant or allowed to grow indefinitely, and parameters have been varied by a factor of ten or more. The purposes of these very severe tests are to find which uncertain parameters are most sensitive, to gain an intuitive understanding of how various parts of the model function, and to expose flaws in the model structure. I should add that we were not trying to establish that the model was either sensitive or insensitive, but that it was sensitive where we would expect the real system to be and *not* sensitive where we would *not* expect the real system to be.

Although we feel that our own test procedures were stringent, probably the most rigorous testing has been done by other groups who disagreed with the methods or conclusions of the model (such as the Sussex Group). They put the model through tests that our own paradigmatic biases would never have suggested to us. Of course, because of those biases, we believe that most of those tests were quite meaningless. They did not alter the policy conclusions we would draw from the model or our confidence in the conclusions. However, the tests did reveal characteristics of the model that we had not fully understood. On the whole, the tests by critics were extremely worthwhile, educational exercises — examples of the process of scientific skepticism at work. We would recommend to model sponsors that every model be fully documented and then given to its strongest ideological enemies for testing.

e. How can one ensure that the model will correspond to reality?

One can be sure that no model will correspond fully to reality. The purpose of a model is to simplify the sysem until it is comprehensible, not to duplicate the system in every detail. Validity testing is a subject of total confusion in the modelling field. It is probably best to admit that validity is a purely subjective concept and that each modelling school has its own unique approach to establishing confidence in its models.

As far as I can tell, each different approach to validation, when used honestly and intelligently, is an informative process indeed, but each approach can also be used sufficiently dishonestly or unintelligently that any model can be 'proved valid'.

f. How can one be assured that the model did correspond to reality?

I assume this question refers to fitting historical data. World3 is initialized in 1900 and runs continuously, without re-initialization, to 2100. Therefore it must duplicate 75 years of history, and over 200 endogenous variables can be compared with their historical counterparts (where data are available). The model matches qualitative, dynamic historic trends quite well, and it even matches numerical values with good accuracy, although we did not emphasize 'tuning' to make this happen. We are not much impressed with the model's ability to match history quantitatively as a criterion in itself; this matching is not difficult to achieve.

g. What is the role and purpose of sensitivity tests?

I believe I have already answered this question. I would like, however, to add one point about sensitivity testing. When we test for the purpose of identifying sensitive uncertain parameters, we define 'sensitive' in a way that does not seem to be commonly shared in the modelling community. We call a parameter sensitive, not when a change in it changes the value of an output number but when it changes the policy conclusions we would draw from the model. Thus many numerical changes in World3 might shift the value of global population in the year 2000 by ±25 per cent. But very few such changes would shift the entire system behaviour from unstable to stable, or would cause policies we have identified as stabilizing to become destabilizing. Because of this pragmatic definition of the word 'sensitive', we label many parameters as insensitive that would be called sensitive by other definitions.

6. Internal organization

a. How should global modelling work be organized?

Our own experience in organizing interdisciplinary modelling work is so limited and chaotic that I would not presume to prescribe operating procedures for others. I can only report on the procedure that finally seemed to work for us.

The model fell naturally into five sectors. After trying unsuccessfully to keep everyone involved in the progress of all five sectors, we finally assigned responsibility for each sector to one or two senior researchers. We carefully specified what information passed from each sector to each other one, and what historic trends in those cross-sectoral variables had to be matched. Then we separated the sectors physically, each sectoral team developing and testing its own separate model, making inputs from other sectors exogenous. Since we had agreed upon the interface conditions, the model worked immediately when the sectors were reconnected. The disadvantage of this approach was that the varying interests, talents, and knowledge of the team members resulted in uneven development of the sectors.

b. What rules and procedures would you lay down to ensure that documentation of the model is kept continuously up to date?

Global models, above all, should be good examples of the modelling art and should be accessible for discussion in a wide variety of arenas. This means that their documentation should be (i) clear to both analysts and laymen, (ii) easily available, and (iii) timely. Not one global model, including our own, has met all three of those documentation conditions; most have met none of them. This lack of documentation is a disgrace. No other supposedly scientific discipline would permit such irresponsible reporting.

The absence of documentation is a result of very understandable and difficult-to-combat characteristics of the modelling field itself. Modellers like to model, not to write. Modellers are not very well trained in model documentation, and many are verbally inarticulate. A model is always evolving, and it seems a waste of time to stop and document something you suspect you are going to change anyway. A really clear documentation is an incredibly tedious job, and it requires an understanding of the model that many modellers in fact do not have. Clients are usually impatient for model results and uninterested in supporting publications. Global models have the added problem of a popular appeal that can generate great pressure to publish general results before technical documentation.

I see no easy way to overcome all these problems. Rigorous standards are needed, rigidly enforced. Perhaps the only way to achieve the necessary discipline is to tie continued funding to regular delivery of acceptable, up-to-date documentation.

Good documentation habits *can* be acquired without bringing modelling progress to a halt. Jay Forrester now imposes a continuous documentation requirement on the group that is developing his national economic model. Every speech or publication that is based on model results, even from interim versions of the model, is backed up by a technical publication available from MIT that gives the exact equations necessary to produce those results. Unfortunately, the non-technical documentation is not kept quite so up to date, and, because of journalistic constraints, it is usually not possible to publish equations along with each article. However, Forrester's past modelling efforts have also set a very high standard for final publication of a model. Somehow, standards at least as high as his should be self-enforced within the entire modelling community.

7. Relations between modeller and user

By trial and error our group has developed a number of guidelines for client relations that we try to follow in our normal, policy-related research projects. They include:

- Find a client who is in a position to implement your results.
- Let the client define the problem; do not try to force a definition on him.
- Listen carefully to the client's ideas; he often understands the system quite well.
- Involve the client regularly in the model's development; if he doesn't understand the model and see his own assumptions about the system in it, he will never implement it.
- Always explain the model with clear language, graphs, and charts; avoid mathematics and professional jargon.
- Have a prototype model running quickly to use as a demonstration and to solicit criticisms and suggestions.
- Meet deadlines.

I suspect that very few of these guidelines apply to global modelling, however. Clients who are really responsible for long-term issues of global scope are almost non-existent. In a sense, the clients for global modelling are all of humanity, or at least that section of the world population that is sufficiently educated and free from poverty to think about the long-term future. All paths to social change do not go through the actions of the power elite; some go through the spread of clear, compelling ideas through the general population.

I have already said that one product of formal models can be ideas, new forms of understanding, and clarification of mental models. The global models already completed have provided such ideas: physical exponential growth can be only a passing phase in human evolution; the pace at which the poor can catch up with the rich in our current system is painfully slow; world food trade patterns effectively reward the rich and batter the poor; it is now physically possible to fill basic human needs. If these ideas were accepted, their 'use' could cause an evolution that could amount to gradual revolution, but the client or source of implementation would never be identifiable.

The Mesarovic/Pestel Model

Source of the response

The response for the Mesarovic/Pestel model was prepared by Professor M.D. Mesarovic, Professor of Engineering, Case Western Reserve University, Cleveland, Ohio, USA.

Professor Mihajlo (Mike) Mesarovic was co-director of the Mesarovic/Pestel Project and co-author of the popular work that describes the results of the project, *Mankind at the Turning Point*. He is also a co-author of most of the numerous technical reports produced by the project during its seven-year life.

Professor Mesarovic's response is primarily an expression of the philosophy that has guided the work of the Mesarovic/Pestel team. He devotes little attention to the 'history' of the project. A strong commitment to implementation of results, which is evidenced by the recent activities of the group, comes through clearly. Mesarovic also places particular emphasis on the incorporation of 'cybernetic' decision mechanisms within the model as a way of dealing with the inherent uncertainties that complicate the task of modelling political and social decision processes.

Comments on the introductory remarks

The comprehensiveness of a model is not due to its globality, but to its long time horizon: the longer the time horizon over which the systems evolution is analysed, the broader and more comprehensive view of the system one must take into account and the more extraneous or exogenous inputs and factors will have to be internalized. There are global models (for example Project LINK* and some sectoral agricultural and/or energy models) that are both short term and limited in scope.

The need to look over a long time horizon is the consequence of the dynamics of the system under consideration. If one were to study the development of a single country (for example Mexico) whose population will, with great certainty, increase (for example from 65 to over 120 million people by the year 2000), the need to consider a thirty-year time horizon is as vital as for the analysis of the global system.

An equally important characteristic of the global system (which, unfortunately, is too often overlooked in the global models) is the structure of the world system as a set of interconnected but autonomous 'self-organized' decision-making subsystems.

The purposes and goals of global modelling

a. What are the main problems (what is the single most important problem) a global model should try to analyse? To what extent can global modelling serve this purpose?

The single most important problem that a global model should try to analyse is the so-called *world problematique* identified by the Club of Rome; namely that mankind *is* entering a new era in its history in which a global society is emerging. Obviously, this society is emerging, not in an organized manner or by 'design' but as an inevitable consequence of a finite globe and the

*Project LINK has as its objective linking national econometric models in order to place national econometric forecasts within a global perspective. The project is directed by Professors Lawrence Klein and Bert Hickman and is located at the University of Pennsylvania's Wharton School of Business Administration.

interdependence of the parts within the global system. The basic question is: how should human affairs be managed so that progress — even if at a slower pace — can continue during the period of transition?

A word of caution is needed in talking about global problems. A problem is obviously global if it affects all the inhabitants of the world; the solution to such problems, however, need not be global in the sense of a centrally directed action. For example, a global energy shortage would probably affect everyone and the solution to the energy problem could be expressed in terms of a need for global investment (for example 40 trillion dollars over the next fifty years) to construct the energy supply system needed. The realistic and meaningful solution, however, would have to relate to where the investment funds will be generated and where they should be used. A new high-technology energy supply system built in Europe would be of little value to people in Black Africa or South Asia. Similar remarks apply to the problem of food shortages; the question is not whether a sufficient amount of food can be produced to meet 'average' global requirements, but rather whether it can be made available where it is needed and to the extent that it is necessary.

b. What are the specific features of goal setting in a global modelling effort? How should normative aspects interrelate with descriptive aspects? What should be the predictive value of a global model? Under what conditions? If no absolute prediction is possible, how should the model be put to use? What kind of a 'global goal function' could be conceived, or what other possibilities of representation of goal-seeking behaviour do you see in global modelling?

Goal setting cannot be envisioned on the global level but only on the level of the social and economic entities that constitute the interwoven parts of the global system. Furthermore, meaningful goals at this level can be neither numerical nor single valued. In other words, a goal is a qualitative concept that can best be expressed as a set of at least partially conflicting 'objectives'. By an 'objective' I mean something that can be pursued more explicitly and for which a more specific strategy can be defined. For example, the overall goal for a nation could be 'prosperity', but, in order to become operational, it would have to be translated into specific objectives, such as the elimination of poverty, increases in leisure time, etc.

Before leaving the topic of goals, I wish to emphasize strongly the important interrelations between the normative and descriptive aspects of models.

If a normative model is to have any credibility (and therefore usefulness) it must be sufficiently 'descriptive'. In other words, if the model is based on a utopian view of reality, which ignores real-world constraints, the normative aspects of the model, no matter how laudable, become meaningless.

c. How would you proceed if you had (i) limited resources and (ii) practically unlimited resources to spend on global modelling work?

With respect to the resources needed to build a global model, one can state flatly that, no matter how plentiful the resources, one can envision a global model that would be complex enough to render them insufficient for the task of constructing it. On the other hand, Jay Forrester has shown that a small model, essentially developed by one man, can be both useful and influential. Therefore, there is a full spectrum of tasks and needed resources; not unlike modelling in other areas (modelling of biological systems, chemical processes, etc.). The resources ought to be commensurate with the objective and task and could be anywhere from one person up to fifty people. Therefore one can envisage a full spectrum of global models from the very simple to the very complex. For each model, resources and staffing should be commensurate with the objectives.

d. What services might (will) global modelling be able to render in the future? What services not?

Global modelling can provide at least the following 'services':

(i) Help decision makers who face the challenge of acting 'in the interests of mankind' or against it — with future generations in mind or ignoring them — to make decisions that are consistent and compatible with the new reality of an interdependent world in which anticipatory actions are mandatory.
(ii) Help in the public domain as convenient vehicles to debate the interrelations of issues and the interdependence of nations.
(iii) Help to bridge the gap between the professional experts and non-professional users of the model's results, if the system is designed as interactive and a 'tool' for policy analysis and planning.

2. Methodology

Given the purposes spelled out under section 1, what are the consequences with respect to methodology? For model structure? For mode of model use? How far do certain methodologies reflect/determine certain world views?

To meet the objectives discussed above, the model constructed has to be flexible. This requires, in particular, that it have two features:

(a) It has to be *modular,* so that within a highly aggregated model on the global level, particular regional or sectoral concerns can be modelled in detail. A good example is the feature in our model that allows us to disaggregate the Latin American and Middle East Regions into individual nations in order to explore alternative national policies within a global context. In addition, we have recently developed a much more detailed economic model, focusing on technological issues and unemployment, for the West German government. This model can also replace a much more aggregated one in the overall system of global models.

(b) The model also should be *interactive* so that it is very easy for policy makers to analyse a full spectrum of scenarios. Because of the importance we attach to this feature, we prefer to refer to the output of our project as a *policy-assessment tool* (or, sometimes, a scenario-analysis tool), rather than a model.

With respect to how the global system is represented within the model structure, the following aspects are important:

(a) The model should reflect the diversity and interrelatedness of the world system at the same time. One of the principal challenges in global modelling is how to represent the interrelations among different parts of the world and among different issues.
(b) The model should be, at the same time, multidisciplinary and interdisciplinary; by this we mean that each set of phenomena considered must be represented consistently with the body of knowledge in the relevant scientific disciplines, while at the same time all submodels — acceptable in their own right within given disciplines — will have to be integrated in a total system as they most certainly are in reality.
(c) The model should at the same time be sufficiently specific to be useful in practical policy analysis and sufficiently comprehensive that one can relate it to an intuitive understanding of the situation.

The hierarchical systems approach, which has guided the development of our model, is particularly helpful in achieving this objective.
(d) In order to take into account the uncertainty involved in representing decision-making processes — certainly this is a vital part of the total system — the model ought to be constructed as a 'cybernetic' system. By this we mean that decision-making processes within the system — such as resource allocation or selection of options among different alternatives — are represented explicitly as such.

By contrast, many models (differential/difference models and input-output models, for example) are strictly deterministic. They assume erroneously that the paradigm that will be useful for describing complex social-economic processes is similar to that of Newtonian mechanics. In the alternative approach we propose, scenarios can be generated by varying parameters that alter search or decision strategies rather than functional forms or initial conditions.

3. Actors, policy variables

Who are (a) the actors in the model whose behaviour is endogenously modelled, (b) the actors steering the model, and how are they represented? How does the model handle policy variables?

For a proper consideration of various and actors and policy variables within the system we consider – as already implied in response to earlier questions – two aspects to be important: (i) the interactive use of the system and (ii) the cybernetic representation of the decision processes within the model.

An additional comment on the 'cybernetic' representation of actors may be in order. A goal-seeking system can be represented either in what is called in biology a structural or functional manner. For example, a structural representation of an organ is its description in terms of chemistry, physics, etc. A functional representation of an organ is in terms of how it fits into an organism and what function it performs (rather than what components it consists of). When we are talking about actors, decision processes, and strategies in a world model, we are certainly talking about functional representation. In other words, the search for optimization or equilibrium procedures – or whatever other processes one uses in the model to represent the goal-seeking behaviour – is only an equivalent representation of the system functions in the real-life system; one does not try to mimic in a one-to-one manner the detailed structure of the decision process. For example, the response of a national system to changing global conditions would be determined in terms of some objectives and strategies that are the net results of an interplay of many national forces, such as unions, industrial leaders, political and economic groups, etc. Finally, not all cybernetic feedback loops will have to be closed within the model. As already emphasized, interactive modelling ought to be used in this area too.

4. Structural aspects

a. Regarding past modeling efforts, could one have arrived at the same result (i) with a smaller model or (ii) with a different model (approach)?

Have you developed a formalism by which to decide what to leave out of a model?

Have you made any effort, after the model was constructed, to compress and simplify its essence in a form that would allow for understanding rather than simply complex explanation? What insights does the model yield that would not have been available through other means of analysis?

The question of the size of the model is related not only to the outcome of any particular run (or a set of runs) but also to the credibility of the results. For example, taking any trajectory in time, a dynamic model can be built to reproduce such a time trend. However, the question is whether such an outcome can be justified and defended as reflecting the actual processes in reality.

b. How is consistency among regions guaranteed?

The consistency among regions is guaranteed by appropriate representation of interregional relationships.

c. How does the model represent the fact that actors adapt to changes. . . ?

This has already been discussed in my response to section 3.

d. Should a model be set up as a one-time venture or should it be deliberately designed in a modular fashion, flexible enough to adapt to new insights and/or policy questions. . . ?

This has been discussed in detail in my response to the question on methodology. I believe strongly in a flexible, modular approach.

5. Testing the model

a. Which aspects of your model do you have much confidence in? Which aspects do you have less confidence in?. . . Has your model produced sensible results when subjected to noise and larger disturbances?

The question of confidence in the model (or any part of it) is intimately related with its use. With respect to producing 'correct' results, within a reliable order of magnitude (that is in reference to errors that might accumulate within the system by running the model over a twenty- to thirty-year period of time), we have most confidence in the population model. This is not due to our special skill in modelling, but rather due to the system itself, which has time constants that require many years to accumulate appreciable errors. The least certain assumptions in the model are most certainly those related to the values, decisions, and policies. For example, what will the investment policy be in a developing country regarding the need for energy development, resources, agriculture, or industrialization? What are the priorities that any particular region or nation might assign to the defense budget, vis-à-vis education, etc.

Our model has produced satisfactory results in the face of a variety of unusual disturbances. For example, we have subjected the model to random-process impacts representing weather variation and its impact on agricultural yield.

b. How can one be ensured that the model will correspond to reality (validation)?

c. How can one be ensured that the model did correspond to reality (calibration)?

Validation has two strictly different aspects:

(i) A model can and should be validated against past data by initializing the model and letting it run over appropriate past time periods.

(ii) Another type of validation refers to the structure and the appropriateness of the model structure for extrapolation into the future. By a proper selection of parameters ('tuning') one can fit a number of different models to past data. However, the appropriateness of using any one of these models for assessing the future is not necessarily determined by how well historical data are reproduced. It is more important for the structure of the model to be properly selected, so that alternative future paths of global development can be reasoned out and logically justified.

6. Internal organization

a. How should global modelling work be organized? How can consistency between subgroups be secured?

The organization of a global modelling team should be based on a common conceptual foundation. In our group, this foundation was provided by the multilevel hierarchical systems approach. The emphasis of this approach on modularity allowed different submodels to reflect the state of the art within relevant scientific disciplines, while at the same time facilitating the integration of submodels into an overall structure. Also it was helpful in minimizing irrelevant conflicts about the appropriate paradigms to be used *within* submodels.

b. How should global modelling groups cooperate in the future?

The question of how world modelling groups should cooperate is not necessarily different from the question of how scientists in other disciplines ought to cooperate.

The challenge for global modellers is not how to communicate to each other but how to communicate to users and even how to identify the user. In too many areas of systems research the researchers spend more time communicating to each other than assuring the practical impact of the results – without which the entire effort is futile.

7. Relations between modeller and user

The relations between the modeller and user should be as intimate as possible, regardless of whether we are talking about a global model or local model. Whenever we are talking about using a model for policy analysis, it is important to receive inputs from users as soon as possible in the model-development process.

The Bariloche Model

Source of the response

The response dealing with the Bariloche model was prepared by the ASDELA Group at the Candido Mendes University, Rio de Janeiro, Brazil: Guillermo A. Albizuri, Irene Loiseau, Carlos A. Ruiz, Neantro Saavedra Rivano, Hugo D. Scolnik, and Jonas Zoninsein. (As only one of the authors was part of the original team that constructed the Bariloche model, the response does not necessarily reflect the opinion of the rest of the authors of that model.)

The response prepared by the ASDELA Group was one of the most thoughtful and complete received. It addresses each issue raised in the questionnaire in considerable detail. Moreover, the group also included a detailed and deep discussion of their philosophical positions regarding modelling.

This group, as well as the Bariloche model, is distinguished particularly by its emphasis on ideological issues. In their judgement, all of the major issues related to model development and use are strongly ideological. Another view different from that of some other groups is the belief that 'no explicit client should be identified'. The group is quite willing, however, to have its model used by those who agree with its normative and technical content.

Introduction

Before answering in a sequential form the proposed questions, it seems convenient to expose some general considerations about global models shared by the members of our group.

In one form or another each question is related to the following basic theoretical notions, and therefore each of our answers will be referred to them in an explicit or implicit way.

1. Global models and reality

Basically, a global model is a structured discourse (we shall develop below the idea that it is a complex discourse with multiple levels). This means that it should never be understood as a 'reflection', or a 'synthesis' of reality. Rigorously speaking, a global model does not necessarily discourse about reality.

In every model, we can always distinguish two main parts: the theoretical model and the formal model (more subtly, a greater number of parts could be identified, theorization and formalization being carried out in several steps; however, it will be sufficient to our purposes to distinguish these two parts). Of course, both the theoretical and formal models are composed of a certain number of submodels.

The theoretical model is structured from the theoretical framework in which the global model is imbedded, through an 'ad hoc' theorizing effort. In the next step, a homology is *assumed* to exist between the theoretical model and reality. It is therefore, a discourse about reality.

Starting from the theoretical model, the formal one is built up through a formalization and quantification process. Now, another homology is assumed to exist between the theoretical and the formal models. Thus, the formal model is a discourse about the theoretical model.

The previous distinction is important for understanding the real meaning ofa hypothesis in each of the models.

In the theoretical model, any hypothesis relates to the elements of the model, and its validation or refutation has a meaning within the model itself. The validation or refutation of this hypothesis in terms of empirical reality is only meaningful if we can previously corroborate the assumption regarding its homology with the theoretical model.

Similarly, a hypothesis implemented in the formal model has a meaning within this model, and its incidence on the theoretical model depends on the validity of the assumed homology between the formal and the theoretical model.

Unfortunately, in the known global models the structured linkages between the theoretical and formal models are not always clear, nor is the relation between them and empirical reality. Very often it is difficult to identify the limits of the theoretical model and to find out the transformation rules that may allow the hypothesis of homology with the formal model to hold.

On the other hand, the situation usually encountered is that *both* hypotheses of homology are assumed a priori and, even worse, tacitly. It is in this way that, subject to certain conditions, the behaviour of the variables in the formal model may come to be seen as a faithful representation of the real historical process.

This lack of epistemological rigour brings risks not restricted merely to academic fields. For instance, an unaware observer (maybe a user of the model) could confuse mathematical simulation with forecasting the future.

2. *The levels of the structured discourse*

We have mentioned that, as the theoretical model is a discourse about reality, the formal model is a discourse about the theoretical model. Considering now the intrinsic unity of the global model (which does not necessarily imply intrinsic coherence), we can now say that the model constitutes a complex discourse about reality and itself, and that several levels should be distinguished in it.

(a) The model as a descriptive discourse. A model sustaining, implicitly or explicitly, the validity of both hypotheses of homology intends to describe reality. This is the level of descriptive discourse. The hypotheses on the presence or absence at a given moment of those elements or variables considered in the model belong to this level.

(b) The model as an explicative discourse. It is at the explicative level that causal hypotheses are assumed and explicit relationships among variables are postulated. The explicative level is superior to the descriptive level, containing it. At this level the discourse seeks to explain reality and it also intends to explain itself.

(c) The model as a predictive discourse. The predictive discourse includes the descriptive and explicative levels. The validity of a predictive hypothesis depends upon the possibility of obtaining a suitable anticipatory description by using a valid set of explanatory hypotheses.

(d) The model as a metatheoretical discourse. The model is a discourse regarding reality and itself. However, it refers to itself, not only in terms of the content of the theoretical model but also with regard to the totality of its theoretical and methodological structure, the relations among submodels, the transformation rules that allow the formalization, etc. In this sense, it is also a discourse about the theory of models and, therefore, is a metatheoretical discourse.

(e) The model as normative discourse. A global model is, before anything and after all, a normative discourse. In other words, the model, at all levels of its discourse, is structured from an assumed normativity that appears in the final discourse. At the description level, the normativity imposes conditions about the selection of variables and the assumptions of relevance and pertinence. At the explicative level, the normativity is implicit in the hypothesis regarding causality criteria and relations among variables. At the predictive level, the normativity limits the spectrum of expected trajectories. At the metatheoretical level, the global conception of the model and the modelling process is imbued with normativity. Finally, a completed global model always offers, first of all, a normative message.

3. The relations among the levels of the discourse

To have pointed out analytically the different levels of the discourse in which a model can be understood should not lead us to lose sight of its unity: after all, it is a single discourse. However, it is important to keep clear the differences among the levels. This is so because each level has its own scope of problems and its own spectrum of dominance and subordination relations with the other levels.

It is necessary to point out that, although any model can be understood at these levels of discourse, not every model explicates them in a clear way. We believe that, in the present state of the art, more remains tacit than is developed explicitly. This is specially true at the normative and metatheoretical levels. Just as the validity of the explicative level is conditioned by that of the descriptive level, the validity of the predictive discourse is conditioned by both. From the fact that the validity of a level of the discourse is conditioned by the validity of the levels below it, it does not follow that the validity of the lower level implies the validity of the higher level. But it does follow, for instance, that it is not feasible to obtain a suitable explanation from unsuitable descriptions, etc.

4. The danger of confusing the levels of the discourse

Every discourse of multiple levels easily becomes fallacious when the level of multiplicity is not clear enough. The most obvious fallacy is to assume the validity of a certain level of the discourse from the validity of other levels. For instance, a good fit may serve to validate the explicative level of the discourse and the aspects of the metatheoretical level related to the descriptive and explicative ones. But it says nothing, or almost nothing, about the validity of the descriptive level (which basically depends upon the homological hypothesis in its synchronic dimension), or even less about the validity of the predictive level. Likewise, a complex and even elaborate articulation of components of a model (part of the metatheoretical discourse) cannot be used for validating the descriptive, explicative, predictive, or normative aspects of it.

1. The purposes and goals of global modelling

Ideally, a global model should deal with all the problems relevant to mankind. Unfortunately, the present state of the art only allows us to study a minimal number of the important problems. To identify the 'single most important problem' is a very delicate subject. In principle, no relevant problem is isolated from others. Therefore, it would be more correct to talk about 'problematiques' instead of single problems.

On the other hand, a hierarchical ordering of the different problematiques according to their relative importance is something that arises from the implicit or explicit normativity of each model.

In our case, the problematique we consider the most important is quality of life, understood in a broad sense. We consider that it reflects the consequences for the individual human being of the social, economic, and political processes.

a. To what extent can global modelling serve this purpose?

Global models are the ablest instruments known to date for trying to corroborate complex theoretical schemes (like the quality of life).

One reason for this is that their global character enables them to reflect — with varying adequacy, of course — such complexity in the formalization process in a better way than traditional methodologies. Nevertheless, it is important to remark that global models are still very far from being able to deal with all the theoretical complexity of many theoretical models. On the other hand, work in the field of global modelling is useful in several aspects. For instance, the elaboration, discussion, cricitism, and self-criticism of global models may:

(i) help discover new facets of the problems under study, which may enrich their theory; (ii) be an important training in interdisciplinary creativity and research; (iii) contribute to widening the scope of sectoral analysis.

b. What are the specific features of goal setting in a global modelling effort?

To be able to define goals simultaneously in comprehensive terms and in several fundamental dimensions.

c. How should normative aspects interrelate with descriptive aspects?

The descriptive aspects of a model are a direct consequence of its normative aspects, be the latter explicit or not. They intervene in: (i) the inclusion and/or exclusion of variables; (ii) the assumptions about how these are linked; (iii) the mechanisms of their dynamic evolution (including the excluded variables). In every model, the elements coming directly from the explicit normative aspects are more likely to be subject to criticism and self-criticism than those arising from the implicit normativity in the model. Thus, in order to make the model more transparent, it is necessary to expose as much as possible its normative content. (However, according to our introduction, let us remember that descriptive and normative aspects are not the only ones to be considered in a model. Explicative, predictive, and metatheoretical aspects are also important matters to think about.)

d. What should the predictive value of a global model be?

It makes no sense to talk about the predictive value of a global model because global (like other) models cannot in principle predict anything at all, except in very limited aspects and for extremely short time horizons. This is so, because we have to assume that the variables not included in the model remain constant, are irrelevant, or vary in such a way as to enable us to consider them irrelevant. The model as a predictive discourse becomes meaningful within the field of the internal logic of the whole model. In other words, it is important because of its metatheoretical and normative linkages with the other levels of the discourse. But it is in no way possible to talk about predicting reality.

In our opinion, a model can only be used for the future as a normative discourse. In other words, the descriptive dimension of a model only allows us to study what would happen *if* certain hypotheses are valid.

e. What kind of 'global goal function' could be conceived or what other possibilities of representation of goal-seeking behaviour do you see in global modelling?

The definition of a goal function depends upon the normative values of those who built up the model or, eventually, of its users. For instance, some people believe that to maximize economic growth would solve most problems, while others advocate the necessity of its halting, etc. Therefore, it does not seem feasible to define a global goal function enjoying a universal consensus.

What is desirable for global models is to be flexible enough to allow the alternative use of different objective functions within the limitations imposed by their normative frameworks. In this sense, it is necessary to point out that no model is neutral from the valuative viewpoint. Nor can its value system be totally altered, even if the goals and values of different users are totally different.

It is much more realistic to assume that different actors will have different objective functions and that the dynamics of the system will result from their (often conflicting) interactions.

f. How would you proceed if you had (i) limited resources or (ii) (practically) unlimited resources to spend on global modelling work?

The main practical problem encountered when working in global modelling is the constitution of an interdisciplinary team sharing at least a basic common language, so as to allow truly interdisciplinary research. Otherwise, the danger exists that the final product may turn out to be a simple aggregate of sectoral models lacking an integrated vision of the global problems.

Of course, the larger the number of team members, the greater the difficulty of achieving this task. However, since we must conduct deep sectoral studies, a limited team can only obtain a global model through an oversimplification of the problematique under study. Therefore, intellectual integration for the global approach and detailed sectoral studies are, in principle, conflicting goals. A way to tackle this problem is to create an interdisciplinary and coherent nucleus, which should commend, after defining the characteristics of the global model, sectoral studies to other groups. This means to ask for the construction of submodels with specified inputs and outputs, and for given regions. Our experience shows that to commend submodels to isolated groups brings the danger that they can use variables which are neither plausible of being formalized in their dynamic behaviour nor reasonably definable in an exogenous way.

As a conclusion, if only limited resources exist, the approach would be to create a multidisciplinary group whose members should be carefully chosen (to have a proven capacity of interdisciplinary work should be more important than to have a big 'sectoral' capacity). If the available resources are bigger than those required by the central team, they could be employed for financing well-defined sectoral studies, following a hierarchical order of priorities carefully decided by the multidisciplinary team.

g. What services might (will) global modelling be able to render in the future? What services not?

The services that ideally a good global model should render are to offer an integrated structure for: (i) testing hypotheses; (ii) calculating 'optimal' trajectories for reaching given goals; (iii) studying the possible effects of international cooperation; and (iv) detecting possible bottlenecks, etc.

A global model should be freely available for continuous improvement, change, testing, etc., by different groups of scientists. Its structure should be as flexible as possible in order to facilitate testing different hypotheses, submodels, etc.

Obviously, no model is satisfactory in every aspect, and, therefore, the most we can aim at is to get a model allowing for gradual improvement according to increased knowledge in each of the areas it covers.

The services it should not try to render are those of intending to 'guess' the future or to try to give 'magic' recipes for solving the problems of a country or region.

2. Methodology

a. Given the purposes of the model, what are the consequences with respect to methodology? For model structure? For mode of model use?

The methodological consequences have been sketched above. Since there is no universal agreement on methodological approaches (for example system dynamics, optimal control, numerical simulation, etc.), it would be important to define standards for testing and validating models and submodels, like those used for validating and certifying algorithms.

The structural design should be oriented towards modularity and portability (like in the design of numerical software libraries), trying to avoid the situation, for instance, where the model would only be able to function with certain submodels of very special characteristics. The structure should also be flexible enough to allow using the model as a goal-seeking instrument (we are here trying to avoid the use of the words 'projective' and 'normative', since a model that pretends to project into the future is in fact normative), the alternative definition of objective functions and scenarios, etc.

b. How far do certain methodologies reflect/determine certain world views?

A chosen methodology necessarily implies a vision of the world. For instance, a model like World3 is such that the initial conditions determine the 'future' (it corresponds to the integration of a system of differential equations with initial conditions). This approach assumes that the world will continue to be like it is, an inertial body subject to a dynamics defined once and for all. The possibility of introducing some exogenous changes at certain moments in time does not basically alter this vision. On the other hand, a model like Bariloche's, which uses optimal control, assumes the feasibility of defining goal functions agreed upon by the whole society and, furthermore, that it will be also possible to achieve basic agreements allowing the practical implementation of the optimal policies.

Again, the possibility of varying objective functions and/or constraints does not basically alter this viewpoint. When running the model in a 'projective' way, the considerations made for World3 are also valid.

For judging what is reasonable and what is not, we have to take into account the goals of the model. If one wants to 'prove' that the present trends lead to a global catastrophe, then a projective method seems suitable. If, on the contrary, one wants to show the feasibility of solving the fundamental problems through deep sociopolitical changes, then an optimization approach is adequate. If one is looking for a model to be applied to present reality, then it would be reasonable to optimize in certain areas (for example energy planning) and to admit exogenous scenarios in others (for example political variables).

Finally, we believe it is necessary to move towards modelling structural changes.

c. *How should aspects be handled that one knows to be important but lack data or knowledge of relations (for example environment or 'the human factor')?*

With respect to lack of data and/or knowledge of relations, the essential point is to admit that, with or without models, every society has to take decisions involving these factors. In this sense, besides experts' judgements for defining plausible relations, it is important to study deeply the model sensitivity with regard to these variables and to present a spectrum of runs under different hypotheses, covering the range of their variation. However, this is far from simple, because of the computational barriers encountered when dealing with complex models.

d. *Many modellers have used cross-sectional, static data rather uncritically to estimate what are essentially longitudinal, dynamic relations. Do you perceive this as a major problem?*

The problem of using relations derived from cross-sectional data in a dynamic sense is a serious one that we are forced to deal with when modelling the international system. It is obvious that it would be desirable to have suitable time series, but they do not exist for many variables. Moreover, a good part of the published data are, to say the least, dubious. It seems to us that the only feasible approach for the time being is to use functional relations estimated by means of robust statistical techniques, representing in an adequate way the phenomena under study for all countries. It is also important to employ weights reflecting as much as possible the relative reliability of the data. For instance, to use demographic functions giving good results for a wide range of countries is, in our opinion, a more reliable approach than to accept nearly arbitrary exogenous dynamic variations of these variables.

An important goal of global modellers is to discuss with the United Nations agencies gathering data the way of doing so. For instance, it does not seem to be very costly to gather dynamic information in the censuses (for example, not only to ask if a person is matriculated in the educational system but also if he was the year before; the same applies with other variables like employment, etc). This would allow us to apply other mathematical techniques like Markov processes.

The use of cross-sectional data presents inconveniences, but, once again, their input depends on the theoretical framework. For instance, if we accept the hypothesis that changes in calorie and protein consumption, education, etc. will affect life expectancy in a similar form in different countries, then a function should exist representing this phenomenon. In this case, data corresponding to different countries can be thought of as points on the surface defined by the function and, therefore, they can be used for estimating its parameters. This does not mean we accept that the so-called underdeveloped countries should follow 'development' paths similar to those of the industrialized world. What we believe is that, for certain variables for which theoretical considerations allow us to postulate that the effects on them of other indicators will be alike in different countries, cross-sectional data may be used in the sense described above. Another consideration is that data of all countries do not constitute a classical statistical sample of the corresponding space, because it is the whole space. Therefore, the usual statistical tests are meaningless. This does not mean one should forget data errors, but the preceding comments are important for the discussion.

The cross-sectional approach seems to be the only operational one in most cases. However, one should take into account the warning that relations may be time-dependent and that excluded variables may change in such a way as to affect the results seriously.

e. How can non-material needs be represented?

The relation between 'need' and 'satisfaction' appears in a concrete social historical context. It always implies the existence of a material base and of a significance universe, within which such a relation becomes meaningful.

As far as a need (or a set of needs) is recognized as such, this recognition entails the possibility of being expressed as a satisfaction demand, addressed to the social structure.

Material and cultural elements are present in this demand, as well as in the structural response. Who experiences a need is, at the same time, a material and a cultural entity. And so is the social structure into which he is inserted. Even the more obvious biological needs (as those directly related to physical survival) cannot be understood outside the culturally significant context in which they appear. This is so, among other reasons, because the subjective manifestation of these needs, as well as the social distribution of their satisfaction, depends on a great number of elements that belong to levels superior to the biological one. Moreover, the form in which the social satisfaction of these needs is achieved has consequences that go beyond the individual metabolic processes.

On the other hand, the more specifically cultural or affective needs (for example those related to artistic creation and recreation) are strongly linked to and conditioned by the material situation of individuals. Furthermore, their satisfaction depends, among other factors, on certain quantitative and qualitative availabilities of material resources.

Of course, it is possible to differentiate and to classify needs. Among the variables to be taken into account, one may consider the degree of urgency of a need, and the social consequences, in the short and long term, of its satisfaction and non-satisfaction, the historical possibilities of meeting the need, etc. But, what cannot be done is to separate rigidly the needs between 'material' and 'non-material'. To do so would entail a philosophical definition, clearly tied to the dualistic tradition, with which we disagree.

With respect to the treatment of needs in global models, it is obvious that it is much easier to formalize the ones more directly related to physical survival than others not so explicitly 'biological'. It is so because the degree of satisfaction of physical needs is often identified with the consumption of certain elements (for example calories, proteins, etc.), and because quantitative studies about minimal standards of satisfaction for these needs do exist. This is not the case for the others.

On the other hand, even if adequate modes are found for formalizing other needs, in no case is it possible to attempt to represent the total set of human needs in a model exhaustively. It is so, among other reasons, because this set is historically variable.

Therefore, perhaps a valid way to represent needs in global models would be to characterize different social-political structures according to their material capabilities and their normative orientation with respect to the satisfaction of human needs considered globally. The needs for which suitable formalization and quantification criteria could be achieved would play the role of indicators in such a characterization. Of course, the pertinent theoretical criteria must always be formulated conveniently.

f. How can physical constraints be best treated?

What is important to point out here is that physical constraints are nothing but data about nature, *such as they are understood by man, at a certain determined historical moment*. What has relevance for global modelling is the social use of nature by human groups. In this sense, what appears again as a valid strategy is the characterization of social structures according their normative orientation to the use of nature, given certain historical levels of technical skill and their corresponding physical constraints.

g. *How should the price system be taken care of?*

When any reference to the price system is to be made in a model explicitly, the use of price indexes for products, groups of products, sectors, etc., constitutes the most direct way or representing it.

In a simulation model it may be useful to look at the mechanism of price formation. If only 'economic aspects' are taken into account, this would lead us to analyse the institutional characteristics of the markets and the normally antagonistic interaction among economic agents that will ultimately set the prices. The globality of a model should also lead to considering non-economic (for example sociopolitical) factors relevant to the process of price formation. Thus, prices would appear as the result of confrontration and conflicts among different social sectors, rather than as an equilibrium solution.

3. Actors, policy variables

a. *Who are (i) the actors in the model whose behaviour is endogenously modelled, (ii) the actors steering the model, and how are they represented?*

A few words on the notion of actor for a global system seem necessary. The notion of 'actor' belongs to the theoretical framework of a model. In fact, while some well-established theories make an essential use of this notion, others can be formulated without an explicit reference to it.

The theoretical background of a model is not always made explicit, and for most, if not all, of the known global models this explicitness, if it exists, lacks the necessary rigour. If in the theoretical framework there is not a meaningful notion of actors, every effort to include them in the formal model or in the computer program will add nothing to our previous knowledge.

The wording of the questions in section 3 suggests that some assumptions were implicit. Namely, to answer these questions requires full acceptance of the following ideas:

(1) Every model has actors.
(2) These actors (a) can be endogenously modelled, (b) may steer the model, (c) have goals, (d) interact, (e) adapt to changes (from section 4).

All this obviously responds to a very restricted class of sociological theories (see also the last comments in section 2 of the Appendix to our response).

If we accept for operational purposes the definition of 'actors' as the elements of a model that can choose among different 'policy' options, then the Bariloche model has four actors, namely the four regions considered. These are the only actors whose behaviour is endogenously modelled, and their interactions are described in terms of international trade and economic aid.

b. *How does the model handle policy variables?*

The policy variables are handled in the model by means of a goal function and constraints. The policies do not remain constant in the model because they depend on the situation of the relevant variables.

For instance, if the satisfaction of one of the basic needs is achieved and the others lag behind, the economic strategy is changed. When facing variations of international trade, the model changes strategies in order to adapt itself to the new situation, etc. If the goal function is changed, then the whole behaviour is different.

The user defines a certain situation by means of constraints and goals (for example bounds on the sectoral rates of variation of capital and labour, urbanization rates, etc.) and the model computes the optimal allocation of resources on a year-by-year basis for reaching these goals.

This approach to changing policy variables according to defined targets derives from the explicit normativity of the model. In other words, it comes from the assumption that goals may be defined so as to be agreed upon by the majority of the population. Therefore, it is also assumed that optimal policies (taking into account a number of economical, physical, educational, etc. constraints) for achieving those goals can be implemented.

We do not think problems can simply vanish from one day to another. We know that new policies require drastic and profound changes. If they cannot be carried out in a certain society, then alternative options can be studied with the model by introducing additional constraints. Of course, this can be done within the boundaries defined by the normative framework of the model.

It is important to remember, however, that here we are talking about how the Bariloche model was built. We are not exposing general ideas about how models should be constructed. Surely, the development of global modelling as a scientific practice will provide more complex and fruitful modes of handling policy variables.

c. *What policy options can the model test?*

The model can test the policy options that can be represented in terms of the model's variables (development priorities, educational and housing policies, consumption and investment alternatives, balance of payments, etc.). It cannot simulate, for instance, variables regarding social and/or political participation.

d. How do the model policy options relate to the policy options and choices being made by leaders today?

In order to give an answer to this question, we will suppose that the word 'leaders' is related to the notion of 'power owners'. (If this is not the case, the problem becomes very much more complex.)

The Bariloche model, when simulating policy strategies that optimize the satisfaction of basic human needs, starts from normative asumptions that have little to do with the actual behaviour of most contemporary power owners today.

Nevertheless, in a broad sense, one can say that the model policy options constitute imaginable policy options for any power owner in the aspects connected to developing and planning policies (whether or not these policies are chosen).

e. How many of the policy variables correspond to policy alternatives that political decision makers can actually select, given their power, and how many are of a more abstract, synthetic nature?

It is necessary to point out that we believe the notion of 'decision makers' to be ideological. It is characterized by a non-historical approach to society which ignores the fact that, in a given societal system, political power is the result of complex social-economic relationships among its members. Therefore, it is essential to understand that wielders of political power are not neutral technocrats, fulfilling a functional role for society, but representatives of determined class interests. Consequently, we prefer not to talk about 'decision makers' or 'policy makers', but about 'power owners'. The exercise of power is viewed as not necessarily related to specific institutional and/or political roles.

The question of how many variables correspond to alternative policies that power owners can select was answered before, at least with regard to what governments can do. For instance, the government may choose a development strategy that is not necessarily followed by the private sector. Therefore, only part of the resources may be affected by a policy reorientation.

It is impossible to enumerate the options because, in most cases, there is a continuum of decisions, which can be considered as different policy alternatives.

f. How should political constraints of an institutional kind be tackled methodologically?

Like other political constraints, the institutional constraints can be incorporated into a model as far as their effects on some of the model's variables can be formalized.

g. What do the structural equations depict in terms of goals of actors?

As we mentioned before, the objective functions and constraints in each region (actors in the international system) describe the goals. Also the sociopolitical and economic goals and assumptions of the model are reflected in all the equations describing the productive system, the allocation of resources, the production oriented towards the satisfaction of the basic needs and not regulated by a market economy, etc.

h. How do the actors interact?

This question was answered when actors were defined.

4. Structural aspects

a. Regarding past modelling efforts, could one have arrived at the same results (i) with a smaller model or (ii) with a different model (approach)?

It is clear the results of the 'doomsday models' could have been obtained without models, owing to the exponential growth assumed for the key factors. However, these models did not throw light on the complex mechanisms linking many variables because of their serious technical weaknesses. What they achieved was to call the attention of the general public to the global problems that exist and affect all mankind.

On the other hand, the task of proving that the barriers for achieving the satisfaction of the basic needs on a worldwide scale are more sociopolitical than physical could have been arrived at without a model. But what the Bariloche model gave were some possible trajectories leading to these goals. It also evaluated quantitatively the complex non-linear feedbacks among demography, education, economic growth, food supply, etc. In this sense we do not believe it would have been feasible to get similar results with a smaller model.

It would not have been possible to use another approach. This is so, because, if one wants to prove certain goals are reachable under certain hypotheses, the optimization approach allows one to calculate optimal decisions for this purpose. When using a common simulation model, if the desired goals are not obtained by trying several alternatives, one cannot decide whether the goals are attainable or whether an adequate sequence of decisions has not yet been found. To try detailed alternatives constitutes, in most cases, an impossible task owing to its combinatorial complexity. In this sense, the approach used in the Bariloche model is extremely powerful when compared with the usual techniques.

b. Have you developed a formalism by which to decide what to leave out of a model?

Basically, the criteria for deciding what to leave out of a model are more normative than formal. However, if from theoretical considerations one intends to explain a certain variable as a function of others, formal criteria exist for eliminating independent variables that do not bring additional explanatory power.

In this sense, we implemented criteria of this sort in the multivariate analysis system which, linked to the data bank, allowed us to obtain most of the relationships.

c. Have you made any effort, after the model was constructed, to compress and simplify its essence in a form that would allow for understanding rather than simply complex explanation?

We have made several efforts to explain the model clearly, its hypotheses, etc., without hiding them behind superfluous technicalities (we believe everything can be explained in clear terms to a normally intelligent person). These efforts have been reflected both in the book and in the model manuals written for UNESCO.

d. What insights does the model yield that would not have been available through other means of analysis?

First, the model shows the material feasibility of building up a society in which basic needs are satisfied. In fact, the model gives many insights that do not seem possible to get by other means. For instance, the counterintuitive behaviour of many variables, the influence of different hypotheses on them, etc. One example of a non-trivial consequence was obtained in the application of the model to Brazil, where we found that, disregarding the hypotheses about the external sector, a type of optimal policy exists for a wide spectrum of alternatives. And there are many more examples.

e. How is consistency among regions guaranteed?

We have used, as in Leontief's model, the pool approach, which has inconveniences (like the other known approaches). However, our emphasis on autarchic development gives some additional support to this approach.

Consistency is not guaranteed, although it is not very relevant owing to the assumption that the net balance of payments goes towards equilibrium (it does not mean zero trade). In fact, it is trivial to devise methods for obtaining consistency, but all the known ones are arbitrary and not based on a theoretical framework.

Without any doubt, regional interaction is one of the weakest areas in global modelling. For this reason, the structure and mechanisms of the international system are a research priority for our group, in order to be able to improve the next generation of models.

f. How does the model represent the fact that actors adapt to changes?

The Bariloche model was designed so as to adapt itself to changing situations while trying to reach the given goals. This assumes the invariance of the regions' goals, but not the invariance of their actions (see answer 3).

g. How should changes in structural relations be considered?

The processes of change in structural relations are discussed in conceptual models which have not yet been formalized. One can conceive, however, that, for extreme values of certain variables, the changes in structural relations will result from conflictive processes among the elements of the existing structure. Obviously, those processes may be formalized only if the underlying theoretical model is able to identify the factors leading to possible structural changes and their interaction mechanisms with the elements of the model.

h. How should technological progress/change be represented?

The treatment of technological change is closely related to the representation in the model of the productive process. It is our belief that a truly adequate account of technology and technological change is not possible within the theoretical framework defined by aggregate production functions. On the other hand, the paramount importance of technology in long-term global modelling does not escape us.

While undoubtedly this is a wide field of research in economic analysis, modellers have still available only aggregate production functions, such as those of the Cobb-Douglas and CES types. For instance, in a Cobb-Douglas function (a common case, and the one selected for the Bariloche model) technological change is represented by a trend factor depending exponentially on time. An alternative treatment is this: if we write the Cobb-Douglas function in the form $P = C(K/L)^{a}L$ (where P is production, K is capital, L is labour, and C and a are constants), thus thinking of the productivity of labour as depending on per-worker capital stock, technological change is represented in terms of a.

Finally, if the economic part of the model is sufficiently thorough, it could be feasible to represent technological change in a more realistic way. The coefficients of technological change could then be derived through the use of a complementary set of relations, eventually making the process of technical progress in the model endogenous. This complementary set should take account of the interdependence among technical progress and other important factors of economic growth, such as the level of education and capital goods production and the changes in the pattern of resource allocation, resulting, for instance, from year-to-year optimization, as specified in the Bariloche model. Following this reasoning, we would recommend that other sources of productivity improvement be included in the model and specified through the following variables: the flow of resources devoted to basic and applied research, development and inventive work of processes and products; the flow of payments for transfers of technology from abroad (in the case of underdeveloped countries); the specific supply of higher education, scientists, and engineers; the capability of the labour force, as well as expenditures for professional training; and industrial organization.

i. What other types of structural relations are important in this context?

One of the important components of the structure of the world system is the economic structure. Outstanding structural economic aspects are: (1) the distribution of wealth and property; (2) the relative development of the productive forces throughout the world; (3) the social and economic organization of the productive process; (4) the economics and sociology of distribution, consumption, and demand. It must be said that global modellers have not been much concerned with these aspects, even though they are essential for the proper understanding of the behaviour they want to simulate. In this case, at least, it is patent that no meaningful statement can be made about the long-term behaviour of the system if one does not take the structure into account.

j. Could 'structure sensitivity analysis' help to answer these questions?

It is not clear to us what 'structure sensitivity analysis' (SSA) is. If SSA is understood to mean a topological approach like that of catastrophe theory, we believe that a good sensitivity analysis of dynamic systems offers more powerful results. It is convenient to point out that more 'structural' variables, like links among submodels, can be included in the sensitivity analysis. Our conclusion is that sensitivity analysis not only helps but is essential for studying these questions.

k. Should a model be set up as a one-time venture or should it deliberately be designed in a modular fashion, flexible enough to adapt to new insights and/or new policy questions? How can such flexibility be assured?

As we have said before, a global model should be modular and flexible and should also change with time. Flexibility can be assured through a very careful structural design.

5. Testing the model

a. Which aspects of your model do you have much confidence in? Which aspects do you have less confidence in? Which aspects of your model require the most subjective judgement? Which aspects are the most concrete? What kinds of errors are to be expected (size and frequency)? What assumptions in your model are you least certain about?

The sources of errors and unreliable aspects in a model are obviously the functional relations, the structural relations, and the data. Data can be divided into two categories: (i) data used for estimating parameters in functional relations and (ii) data that enter directly into the model.

Many efforts have been made in the Bariloche model to justify very carefully each equation. In this sense, functions like those linking demographic and socioeconomic variables, agricultural outputs with fertilizers, etc., have been fitted to all available international data using the best non-linear techniques.

In all these cases we tested different functions until we found relations that are satisfactory from qualitative and quantitative viewpoints.

Of course, there are parameters in the model that are liable to subjective judgement, like the maximum amount of arable land, maximum yield, coefficients of technological progress, etc.

Another sort of problem arises from errors derived from the lack of an adequate formalization in some areas. For instance, child mortality is obtained as the solution of a non-linear equation through an elegant mathematical formulation. The pity is that it does not match reality very well. Another example is international trade, for which no satisfactory model is available.

In order to talk about the errors one could expect from the model it would be necessary to compare its results with a future reality, which is senseless within the context of a normative discourse. Hence, what we can do is to compare the model values with historical time series (see point *c* below).

We have not used statistical techniques or a 'fuzzy-set' approach to deal with uncertainty in the parameters for several reasons. One is that, quite often, data values are modified or invented for political reasons, and therefore they are not subject to statistical perturbations in the classical sense. Therefore, we decided to accept as satisfactory functions (or systems of coupled non-linear equations in demography) that give good results for a diversity of countries whose information was judged to be reliable.

The model produces different results when some variables are subject to large disturbances, which is normal and realistic. For instance, to introduce a 'large' perturbation in the technological progress parameters leads to different results, which is also true in real life (see also point *d* below).

b. How can one be ensured that the model will correspond to reality (validation)?

This question was answered when we discussed models as multiple-level discourses.

c. How can one be ensured that the model did correspond to reality (calibration)?

The procedure for calibrating the Bariloche model seems to be the most elaborate among those known up to date. It is briefly described in *Catastrophe or New Society?* and in a more detailed form in the *Handbook of the Latin American World Model* (UNESCO, Paris, 1977). Basically, it consists of calibrating the model's parameters by means of an optimization technique for dynamic systems. The key point is that it was possible to obtain an excellent reproduction of the period 1960-1970 in several variables of different type (natality, life expectancy, population, calorie consumption, GNP, education, etc.). In all cases the relative errors of prediction were less than 5 per cent.

Of course, to be able to reproduce a certain historical period with accuracy is a necessary but not a sufficient condition for validating a model. What it proves is that the model is coherent in the period considered.

d. What is the role and purpose of sensitivity tests?

The trajectory of a dynamic model (either an optimizing or an extrapolatory model) depends upon the initial data, the parameters of the equations, the topological structure, etc. The role of sensitivity analysis is, ideally, to study the conditions under which the model will produce qualitative or quantitative stable results. If a model turns out to be extremely sensitive to parameters that are not known precisely, then its results cannot be relied upon. On the other hand, to detect rather sensitive control parameters is useful for adopting planning policies.

The usual method for studying a model's sensitivity is to modify one parameter at a time, to run the model, and to see what happens. This is a very poor approach, since it is equivalent to assuming that a function of several variables can only vary along the coordinate axes. A non-linear programming approach that gives excellent results is given in H. Scolnik and L. Talavera's paper, 'Computational and mathematical problems in the construction of self-optimizing dynamic models', in *Proceedings of the Second Symposium on Trends in Mathematical Modelling,* Poland, 1974. One application was the analysis of World3, for which the classical way of studying sensitivity did not find 'unstable' combinations of perturbations. However, the optimization approach showed that perturbations of less than 5 per cent in some poorly estimated parameters lead to a continuously growing population until year 2300! In particular, if one accepts World3 as a 'valid' model, then these perturbations give the policy actions that should be taken to avoid the catastrophic future predicted by the standard run of the model.

An extremely important and still open practical problem is the sensitivity analysis of optimizing models like Bariloche. This is because to perform a complete sensitivity analysis by a non-linear programming approach requires many runs of the model, and therefore it is practical only when the required computer time is rather small (say, one minute per run). A model like Bariloche computes the trajectories by solving a discrete-time control problem every year, something demanding a fair amount of computer time. Moreover, the calibration to the historical past depends on the model's parameters. This means that, for each change in a parameter, the model should be calibrated again before running it. As a consequence, the computational complexity of a complete sensitivity analysis is far beyond the capability of present computers. Therefore, only partial sensitivity tests have been made.

It is also necessary to point out that the classical sensitivity theory is based upon assumptions that do not hold in the Bariloche model, and hence it cannot be applied. Another fact to be taken into account is that goal-seeking models have self-correcting capabilities when facing disturbances.

6. Internal organization

a. How should global modelling work be organized? How can consistency between subgroups be secured?

These questions were answered in section 1.

b. How should global modelling groups cooperate in the future?

The cooperation between relatively coherent groups from the normative viewpoint may go as far as the will of their members. Between groups with divergent normative orientations, it does not seem cooperation can go far beyond information exchange, discussion of techniques, constructive mutual criticisms, etc.

c. What rules and procedures would you lay down to ensure that documentation of the model is kept continuously up to date?

We do not feel morally qualified to pontificate on rules and procedures for keeping documentation up to date. Our group (ASDELA) started working with the model after the first project was completed, and we have had a lot of trouble in gathering all the data and documentation produced by the original team. In spite of this fact, we also worked for some time without caring too much about those aspects.

To dedicate part of the time to producing or updating the documentation of the progress of the work may be felt by modellers as a waste of time. Very often, also, results have to be presented on a fixed date, and there is no time to write up the complete documentation.

On the other hand, we know by experience that not to have documentation that clearly reproduces the development of our work can also be a waste of time.

Moreover, still referring to the internal organization of the work, to keep documentation up to date is very important in order to ensure better communication between subgroups and consistency between submodels.

Besides, good documentation plays an important role in the diffusion of the model. We think that, in order for other scientists and modellers to be able to study and criticize the model, all the data sources, the methods by which unknown or uncertain data were estimated, the exact equations used and their origins, and the computer programs must be available.

This procedure should form as much a part of the discipline of global modelling as it is of any scientific practice.

It is also necessary to state the main characteristics and hypotheses of the model clearly enough for them to be comprehensible to a nonspecialized user.

d. What are the best ways of interweaving with the scientific community at large?

First, it is necessary to describe the present state of the art clearly and honestly. Second, it is useful to write up documents specifying by areas the problems faced when building the global models. These documents should include very concrete questions addressed to the scientific community, asking for advice and help.

7. Relations between modeller and user

a. Should an explicit client be identified at an early stage?

We believe global modelling should be a scientific activity trying to help to study, and perhaps to solve, some of the fundamental problems faced by mankind. In this sense, we think global modelling should neither be a commercial activity nor respond to the limited interests of individuals or groups. Therefore, we would prefer to employ the word 'user' instead of 'client.'

The logical consequence is that we think no explicit client should be identified at the early stages. This does not mean that, once the model is built up, somebody agreeing with its normative and technical content may not become a user of it.

In order to avoid further confusion in this field, we propose the following classification for people interested in global models:

Client: *buys* a model, in a traditional commercial sense.
User: *uses* a model, as a source of information and/or orientation in solving his own problems.
Sponsor: *finances* a model team during the model-construction stage.
Reader: *reads* the publications about a model.
Critic: *studies, analyses, and criticizes a model.*
Modeller: *works in constructing* a model.

b. To what extent did clients participate in the work?

If by 'client' it is understood to be those who financed the work, then, in our case, it was the International Development Research Centre of Canada. They did not participate in the work at all, allowing the authors to carry out the project in the way they wanted.

c. What formats to communicate with the users should be developed? What can be done to help the user to avoid misunderstandings, to understand the strengths, weaknesses, etc., of the model?

The essential point in the communication with the users is to explain, very clearly and without mystification, the hypotheses, the virtues, the shortcomings, etc., of the model. If the user is requested to provide exogenous variables for defining scenarios, he should be aware of the implications of each one of his decisions, that is the empirical knowledge about the variables and what the reasons are when they have not been modelled endogenously.

d. Which users need world modelling most? Which are most interested in world modelling?

The users who are most interested in global modelling are the United Nations agencies, some governments, and some academic groups.

e. Do you have recommendations on how to reduce 'overselling'?

If the previous recommendations about relations with the user are followed, then 'overselling' will be reduced automatically. However, this is extremely difficult to achieve, because those who oversell models do not follow these recommendations.

On the other hand, models should not be sold as 'magic' black boxes — instruments of a new form of intellectual colonialism. In this sense, it is necessary to encourage the formation of autonomous groups in several countries which have the capacity to criticize and compare different models and construct their own.

Perhaps we could think about creating an Ethical Committee of Global Modeling to help solve these problems.

Appendix: on the notion of structure

1. Introduction

Several questions related to the structure of the world system and of models of it appear in sections 2 to 5 of the questionnaire. The term 'structure' is a very ambiguous concept, usually meaning different things to different people. We felt some confusion of this kind when reading these questions and also during the Conference. Some people, when talking about structure, mean the structure of a given model. Others mean the structure of the world system. In this last case, there is also not a single meaning. Some people tend, for instance, to identify economic structure with an input-output matrix, others with technical progress, and so on.

As a clarification of some of our answers, we think it is useful to add some words on the concept itself and some other related notions, such as sensitivity, stability, etc.

2. The concept of structure

If we consider a given real system S, such as the world system, there are two principal connotations associated with the term 'the structure of S'. First, any aspect of S deemed to be deep and going beyond the external appearance of the system is often called structural. In this sense, anything in S that explains the specific behaviour of the system pertains to its structure, where 'behaviour' means the observed succession in time of states or positions of S. A second attribute of the structure is that, most of the time, it evolves slowly, as compared to the evolution of the state of the system. This would explain, in part, why the structure is more difficult to identify than the state (or position) of S.

We will not join here the dispute about whether or not the structure of a real system is in itself a real object. At any rate, it should certainly be acknowledged that it is a useful concept, abstracting the most essential aspects of S, many of them unknown. The concept of structure for a model or, in general, a formal system is essentially different. Here we are confronted with a formal, completely precise notion. The structure of the model is the set of its components: variables, functional relations, data sets, and rules of functioning (including, for instance, optimization procedures).

Now, if M is a model for S, the structures of M and S are different. What M does, or is supposed to do, is to imitate the behaviour of S, but this is not accomplished by building a formal system that has a structure similar to that of S. This is why one should be careful to distinguish, when discussing a model, whether or not one is talking about the structure of the model or of that of the real system.

The notion of actor is related to this subject. According to Michiel Keyser in his evaluation of question 3, one can identify two approaches to modelling: the 'empirical' and the 'axiomatic'. The empirical— exemplified, according to Keyser, by the Bariloche and Forrester/Meadows groups — only pretends to simulate the behaviour of the system by using a structure different from the real one. The axiomatic, by introducing the actors explicitly in the design of the model, reproduces the structure of the real system and, consequently, its behaviour. We disagree with this classification and with these denominations strongly; they seem to imply that any global model will reproduce the behaviour of the real system through an adequate reproduction of its structure in the model. In the first place, we must say that we know of no global model with these characteristics. Second, a fallacy is introduced in this classification by the hidden assumption that the structure of a real system can be fully represented in terms of interactions among actors. Nothing warrants this assumption – which certainly corresponds to a particular vision of the world. Finally, pursuing consistently the so-called axiomatic approach would lead to introducing an overwhelming number of actors.

3. Structural change

As we have said, the structure of the real system changes, most of the time, very slowly. However, at some times these changes are abrupt. They correspond to revolutions, crises, breakthroughs, catastrophes, inspirations, or whatever you may call them in their corresponding contexts. These abrupt changes may appear because of accumulations that 'overflow' the present structure, or through the process of resolving conflict among coexisting elements of the system. No existing global model has been able to simulate these changes successfully, and the models that generate changes in the structure of the model do it in a mechanical way (Bariloche, with the transition from its 'projective' to its 'normative' phase, is a good example of this). In fact, in these cases, the rules governing the changes in the model (for example changes in the functional relations, and so on) are themselves part of the structure of the model, so this structure is, in all cases, totally rigid.

Now let us return to the consideration of the structure of the real system. As it happens, with the exception of moments of crisis, the structure of the system is stable. This means that slight perturbations in it will not affect the behaviour of the system significantly. This is just another way of saying that the structure, besides changing slowly, does it in a way that easily goes unnoticed. However, at a moment of crisis, the structure is quite unstable. Small perturbations in it can give rise to many different behaviours (this is the situation alluded to in our answer to the second question in group 4 (c) of the questionnaire). On the other hand, the structure of a model of the real system must be stable, even if it pretends to simulate the process of structural change. Or, using a common terminology, the model should not be too sensitive to changes in its parameters or its functional relations. It must be repeated now and again that an unstable structure for a model is not a necessary condition for a successful simulation of an unstable system. Even if instability is to be simulated, this should be done in a stable way. Indeed, a model whose structure is unstable is worthless.

The MOIRA Model

The source of the response

The response dealing with the MOIRA model was prepared by three persons: J. de Hoogh, Professor of Agricultural Economics at the Free University in Amsterdam and Deputy Director of the Agricultural Economics Research Institute at The Hague, M.A. Keyser of the Economic and Social Institute at the Free University in Amsterdam, and H. Linnemann, Professor of Development Economics and Planning and Director of the Economic and Social Institute at the Free University in Amsterdam.

These three respondents to the questionnaire are the authors of the most important publication on the MOIRA model, and they directed the project from its early beginnings.

In the letter accompanying their response they stressed that they reported on their experiences only in summary form and that a thorough discussion of all of the points raised by the questionnaire would be impossible in a short paper. On the other hand, they felt that a short response might well bring the main points to the fore more clearly than a longer one.

1. The purposes and goals of global modelling

In general

The single most important problem to be analysed in global modelling is the survival of the human race. Formulated more optimistically this is improving the living conditions of world population, especially of the poor. An important element in this is, necessarily, reducing the growth of world population and, perhaps, the gradual reduction of absolute population numbers. Global modelling may serve these purposes to a limited extent by analysing alternative policy options.

A specific feature of goal setting in global modelling is the weakness of international/world policy making and policy instruments. Overall goals may have to be translated into (compatible) national goals. Present goals and policies should, if possible, be included in the description explicitly. The predictive value of a global model is primarily to be found in demonstrating the implications of alternatives (alternative policies and institutions) under conditions pertaining to reality. As no absolute prediction is possible, the model should be put to use most carefully — that is by comparing alternative development paths under well-defined conditions.

The explicit use of a global goal function in which several goals are combined in some weighing procedure is not needed per se and is not necessarily the best approach. In many cases alternative policy results can be presented and compared and valued by policy makers without the use of a formal criterion.

With limited resources, it would be useful to concentrate on one aspect, with rudimentary representation in the model of other aspects, in spite of the obvious and well-known drawbacks of a partial analysis. As the first step in any attempt at global modelling, the existing literature and models would have to be reviewed.

With practically unlimited resources, it would be useful (a) to spend a good deal of time and effort on the issues/areas where traditional disciplines touch each other without really ‘interlocking’, (b) to pay adequate attention to regional and national differentiation of the model, and (c) to aim at continuous monitoring and updating.

The services that global modelling may be able to render in the future are not to be found in absolute prediction but in showing alternative paths and their implications. Continuous adaptation of the conditional forecasts on the basis of additional data and improved formulation of interrelations will be needed. In this way, global modelling may contribute to sound and timely policy making, provided the limitations of any modelling exercise are kept in mind.

The case of MOIRA

The main problems that were analysed with the help of MOIRA can be summarized in two questions:

(a) What can be said about the development of food deficits in poor countries in the next four decades?
(b) To what extent and in what manner do the agricultural policies of the rich countries affect the development of food production and food consumption in the Third World countries?

In goal setting, MOIRA first of all tries to describe the functioning of the 'world food supply system' under the assumption of unchanged behaviour of governments, their policy goals and instruments being incorporated in the model explicitly. After that, this analytical model is used for normative purposes by varying these policy variables in order to analyse the effects of various policy scenarios on the extent and location of hunger in the world during the period under consideration.

As regards the predictive value of MOIRA:

(a) Many uncertainties limit the predictive value of MOIRA simulations, primarily the uncertainties with respect to the exogenous variables (that is population growth, growth rate of non-agricultural income, income distribution, prices of agricultural inputs).
(b) Within these conditional assumptions about crucial external influences, the endogenously generated variables are still subject to a large margin of uncertainty due to the necessarily high level of generalization with respect to the behaviour of the actors in the world food system (farmers, consumers, governments) and with respect to the description of the technical, economic, and institutional environment in which these actors are assumed to operate.
(c) MOIRA, therefore, has only relative value for prognostic purposes. Its usefulness is chiefly in analysing and illustrating the structure of

the complex of technical, economic, and political factors that affects and interrelates agricultural production and food consumption in the various regions of the world.

(d) MOIRA simulations intend primarily to show the cumulative short-term and long-term effects of external influences or political interventions on the direction in which food production and food consumption in different parts of the world will develop.

Modelling work along the lines of MOIRA continues; the project has entered now what may be called its second stage. In trying to improve the model, the following priorities have been established:

(a) Given the limited resources, efforts to increase the effectiveness of the model will concentrate on disaggregating the agricultural sector with respect to country-specific conditions for production and policies, to commodities, to output increasing methods, etc.
(b) As a consequence, the model will grow in complexity with respect to international relations, which call for in-depth study of the equilibrium mechanisms to be depicted in the model.
(c) Balance-of-trade consequences must be shown more explicitly.
(d) A greater variation in international policies ('new economic order') must be depicted; the mode of describing the policies must be more formal.

2. Methodology

The steps to be taken in developing a world model are not basically different from those in any other (quantitative) research project. They may be listed as follows:

(a) Formulate the problem/problems to be analysed in interaction with
(b) A systematic and 'full' description of reality (verbally).
(c) Decide on levels of aggregation and generalization, and degree of closedness.
(d) Try to establish and quantify relationships, theoretically and empirically.
(e) Simulate historical development paths with the model.
(f) Formulate and model alternative institutions and/or policies.
(g) Compare model runs based on alternatives with 'unchanged policy' runs.

A global model will necessarily reflect a certain world view, as it is a description of reality as (subjectively) perceived by the model builder.

In MOIRA we attempted to start from a verbal description of various phenomena as perceived by several experts. Then we attempted to develop models which would fit these descriptions. Finally, from these specific models a general national model was designed to represent the main features of the specific models and be quantifiable. In the process, several interesting relations were dropped mainly because of lack of data (for example on ecology), lack of expert knowledge (for example on the interrelation between economic and demographic development), or lack of time (to study, for example, the non-agricultural sector in more detail).

A few comments on specific issues may be added here:

(a) Cross-section versus time-series data. It is hardly possible to operate at the world level without making use of cross-section analysis. Time series are either too short or the fluctuations in them cannot be explained appropriately by the crude specifications used in global modelling. Some calibration procedures can, of course, see to it that the model also generates the historical path, but cross-section analysis is always a comparatively static type of analysis. Dynamics can only be introduced through arbitrary techniques (a priori specification of weights in distributed lags, etc.). The most serious problem related to cross-section estimation in global models is, however, the problem of aggregation. Here logical errors often slip into global models when a non-linear equation, estimated by a cross-section over countries, is used uncritically at a regional — or even at the world — level!

(b) The issue of the price system cannot be discussed in a few lines. In MOIRA we pursued a general-equilibrium approach.

(c) Physical constraints entered the model as potential arable land and maximum agricultural production, computed on the basis of a substantial amount of agronomic research.

3. Actors, policy variables

For each country MOIRA distinguishes three groups of actors: agricultural producers, food consumers (twelve income classes), and a national government. The behaviour of these actors is described endogenously.

(a) Government tries to maintain income distribution between the agricultural and the non-agricultural sector at a target level through a domestic price policy in food. Such a policy is not a problem-solving policy, but is merely the result of political pressures. The price instruments chosen are of a rather synthetic nature, as all government policies have been translated into these terms.

(b) Producers are income maximizers. They know the shape of the actual production function under normal conditions and take their production decisions accordingly (purchase of inputs).

(c) Migration and consumption are treated as direct reactions and depicted through econometric functions.

At the international level an imaginary supranational government is introduced. Under 'unchanged policies' this international government does not take any actions, while in policy runs these types of action are taken:

(a) Countries are taxed in order to finance an international food aid programme and food aid is provided.
(b) Consumers are 'asked' to limit their food purchases when prices rise above target.
(c) Producers get deficiency payments (payment for taking land out of production) when prices fall below target.

Although the policies considered are rather simple, they are actually being discussed at international negotiations. The goals are not easy to reach, especially because they require a political willingness to introduce new policy instruments and because they may conflict with short-run interests of the nations.

4. Structural aspects

a. Constructing the model

The model is a learning tool for the modeller. All the main conclusions of the model could in principle have been obtained without it on an intuitive basis. The model, however, selects among several intuitively acceptable options, it confronts the ideas with actual data, and, because of its computerization, it makes it possible to find conditions under which the conclusions are false. It would, however, be very sad if thinking about the future would become a specialized field for computer people and econometricians, not because these people would lack imagination but because the development of a serious computer model requires so much time in itself that not much time is left for 'real' thinking.

b. Consistency among regions

In MOIRA supply and demand are balanced. The balancing mechanism is the world market price for food. MOIRA is a multination model and the world market for food links the nations to each other.

c. Adaptation of actors to changes

The model specifications should take care of this. Problems emerge of long-run versus short-run specifications, time-series versus cross-section estimation, and embodied versus disembodied technical progress. Structural change should be embodied as much as possible; otherwise too many miracles may occur.

d. Model flexibility

The first model is bound to be a 'pilot' model. The modular structure is of course, an attractive one, but it can only be set up once the basic structural design is constant. After a few years of group experience, such a design may, however, exist.

5. Testing the model

a. Confidence in the model

The question regarding the confidence (or lack of it) in the model by the model builders themselves is hard to answer. The overall consistency of the model, the general theory behind it (on consumption, migration, etc.) do not seem unacceptable. Results on world price developments do not seem very stable; the exogeneity of non-agricultural output and of the intrasectoral distribution is a serious shortcoming.

b and c. Validation and calibration

Calibration of the models is not a difficult task if one is willing to sacrifice all the degrees of freedom available. Such a procedure, however, yields totally insignificant parameters. Calibration and validation should, therefore, be considered as one problem (for example a maximum-likelihood estimation problem).

d. Sensitivity tests

Sensitivity tests should only be applied to very uncertain parameters; otherwise the curse of dimensionality drowns all conclusions. Moreover, sensitivity analysis is unable to reach conclusions about the validity of the model. Only when a high sensitivity with respect to a given parameter occurs *and* when the modeller feels (or should feel) insecure about the value chosen in the model is it legitimate to reject the hypothesis.

6. Internal organization

In virtually all global modelling work, many skills are required: representatives of the disciplines involved (in MOIRA: agronomy/plant physiology/soil sciences: agricultural economics/development economics/mathematical economics/econometrics), model builders and/or mathematicians, statisticians, programmers, etc. Cooperation between them requires (a) a basic trust in each other's competence and (b) willingness to adapt, if possible, one's approach to the 'requirements' of the other partners. Obviously there should be agreement on the appropriateness of a quantitative description and analysis, and about the type of model to be built.

As said before, the starting point is the development of a verbal picture of reality, on the basis of literature study and expert knowledge. In discussions between the specialists in a certain field and the model builders, relations are formulated and specified. The (mathematical) properties of the assumed relations should be in harmony with the specialists' understanding of reality.

Consistency between the work of subgroups was secured by (a) the will and determination to stay together, (b) the central role played by the main modeller, and (c) regular (say, monthly) meetings among all persons involved. Aspects that turned out to be too difficult to handle were dropped, however; the initial, more ambitious plan to cover all basic needs had to be abandoned, and the persons working on the non-food basic needs prepared a verbal analysis that was published in popular format.

The MOIRA team had to be organized largely on an ad hoc basis, and no guarantee of permanent employment (or at least for, say, five years) could be given. Particularly during the first two years, financing was very insecure and came from 'odd sources'; the financial situation improved when work was well under way and showed promise of successful completion. Apart from the contributions of those having permanent (university) appointments, the MOIRA project required, in just under four years of actual work, an amount of one million Dutch guilders, or about $400,000 to $450,000.

To continue the work on the world food problem along similar lines, a Foundation was established in July 1976 in which two Ministries (Agriculture and Development Cooperation) and two Universities (Free University in Amsterdam and Agricultural University in Wageningen) paticipate. Of the 'non-professorial' staff of the original team, only two staff members form part of the new team, which is headed by a (new) Project Director. Geographically the team is located at the two participating universities.

Cooperation with other global modelling groups has not been very intensive in the case of MOIRA. Exchange of information and other contacts were maintained predominantly at international conferences, especially the Global Modeling Conferences organized by IIASA and some Club of Rome sponsored meetings. Exchange of working documents and reports is, of course, useful. Only later on did more intensive cooperation develop with IIASA's Food and Agriculture Programme, and over the last two years a staff member has been working at IIASA on special assignment.

While work was in progress, documentation of the model was kept up to date as much as possible, though in the final stage (calibration, trial runs, etc.) only those working with the model knew the ins and outs of what happened. It took more time than was anticipated to get the model in its final form fully documented and described; it is felt that in the 'second' project more attention should be paid to the continuous participation of all group members and to the updating of working papers, etc., describing the model.

Interweaving with the scientific community at large is probably best achieved by the usual methods, that is circulation of working papers, publication of (partial) results, participation in seminars and conferences, and (short-term) exchange of staff.

7. Relations between modeller and user

The MOIRA project was a Club of Rome initiative. No specific instructions were given and the funding came from the Dutch government so that scientific independence could be maintained. In any global modelling exercise only funding agencies should be considered that are already aware of the limitations of global models. Then the modeller is not tempted to make golden promises or oversell his results. UN agencies, governments, and universities should be the main clients.

The SARU Model

The source of the response

The response dealing with the SARU model was prepared by Peter C. Roberts of the Department of the Environment, London, United Kingdom.

He has been head of the Department of the Environment's Systems Analysis Research Unit since its creation in 1972. The Unit completed the first phase of its global modelling work in 1976, when SARUM was presented at IIASA's Fourth Global Modeling Symposium. This model was enhanced and extended during the second phase, which ran through 1978, as scenario explorations were provided for the OECD Interfutures project.

Most members of the team who worked on SARUM were available to be consulted during the preparation of the response to the questionnaire. Thus, the response represents the outlook and philosophy that evolved within the team over the years they pondered the issues and worked together.

1. The purposes and goals of global modelling

It is not sensible to define the main problems, and certainly not the single most important problem, for a global model to analyse, because the object of constructing the model is to discover the dimensions of stress in separate areas of conceivable restraint, such as mineral resources, food supply, energy availability, etc. The modellers might start their enterprise with the conviction that a given area represents the main problem, but, on examination of the model's operation, discover that their initial belief was false.

Global modelling can certainly serve to indicate the dimensions of given potential sources of stress.

There is no especial need for goal setting at all in a global modelling effort (except for the rather obvious one of attempting to formulate an accurately representational model of the operation of the global system). However, it is reasonable to ask why global modelling has arisen at this time — presumably because there is the widespread perception of a class of problems not soluble by elementary methods. Furthermore, if no such perception had arisen, no global models would have been attempted. But one should be wary of taking currently perceived problems and using them to erect some quasi permanent 'goals'. Some problems can be 'solved' by merely changing one's point of view and others only by an action on the environment.

If there are some exogenous variables present, then the so-called normative aspects can enter via the choice that is made for the values of the exogenous variables.

A part of the descriptive power in a global model is concerned with placing limits on the directions that the world system could take. This is based on the argument that, irrespective of the desires and choices of individuals or national leaders, the physical constraints will force the development to lie within certain calculable limits. The broad band of potential trajectories could be regarded as predictive.

Absolute prediction is not possible. The model is useful for indicating the probable outcomes of certain specific policies.

It does not seem to be a sensible enterprise to undertake the definition of a 'global goal function', because, as greater understanding is achieved (partly through the construction of good global models), the objectives of mankind themselves change. There may be a contradiction present in the notion of a global goal function if human society continues to manifest a large number of groups with conflicting aims that are resolved through bargains, compromise, or force.

If only limited resources are available, it is probably best to use a model already developed and modify it appropriately. If practically unlimited resources were to be available, then the best practice would be to finance a series of separate, independent groups (each about twenty strong — but containing no more than five modellers) and require of each group that it set its own goals, develop its own structures. In this way a greater variety of approaches is likely to emerge and the better ones will be discovered more quickly.

Global modelling is likely to enable greater understanding of relatively complex systems to be achieved in the future. Global modelling will never enable us to derive precise predictions about the future. It will never tell us what goals we ought to have.

2. Methodology

Just as for other scientific activities, the construction of a global model should be a matter of representing reality accurately. If the method and structure employed by modellers entail goal setting as part of the operation, it will not be possible to obtain an accurate description. Thus, a heavily normative operation must reflect/determine certain world views. Of course, even the construction of a thoroughly descriptive model is necessarily coloured by the world views of the authors. For example, the set of causal mechanisms allowed is restricted to the scientific paradigm for most if not all modellers. (They will, in general, not suppose that a particular event in world history was 'caused' by a Saturn/Uranus conjunction).

If data are sparse, the quantity of uncertainty which is introduced by this very sparseness should be recognized in the model output by bands of uncertainty. If relations are in serious doubt, then it is better to avoid modelling the areas in which this ignorance is profound. For example, many world modellers have avoided inserting explicit fertility mechanisms, because the causal bases for fertility are poor. Thus it is better to handle population increase in an exploratory fashion rather than to insert dubious causal mechanisms, although this does not preclude trying speculative mechanisms in an experimental style.

Modellers will use both cross-sectional and longitudinal data if they are available. If only cross-sectional data are available, then they have to be content with them. It is better to use the limited data available than to make guesses. In some areas, the lack of time-series data is a serious obstacle in establishing good relations.

There are strategies for dealing with the case of sparse data. For example, one can appeal to the contraints that must have been present in order for the present state to have evolved. There may be poor evidence of the constraints, but logically they must have been present.

It is possible for non-material factors in modelling to be handled through an informational measure (bits). Non-material factors like education can be considered simply through the cost of provision, but this does not meet the difficulty of measuring non-material factors in ways that are not merely arbitrary.

The concept of limits to safety in terms of the maximum desirable value of some variable like carbon dioxide concentration can be treated by modelling a world in which there is successively greater and greater awareness of the significance of these danger limits.

Prices, whether in free markets or in planned economies, can be handled in a descriptive fashion. It is better to model a free market and then to introduce interferences and interventions of one sort or another than to start with the planned case.

3. Actors, policy variables

The actors whose behaviour is modelled endogenously are consumers, workers, investors, and managers or resource allocators. Certain governmental functions are also assumed, so that the model can be regarded as having five classes of actors rather than four.

The 'steering' is associated with the manipulation of exogenous variables, and in every case this manipulation is considered to be undertaken in the real world by governmental agencies. It is, of course, quite possible that the governmental action is itself stirred by the pressures exerted from sections of the populace. There are other changes that can occur besides those induced by governments — tastes can alter and new inventions can appear.

The idea of an 'intervention' is not well defined. For example, if a region suffers a disastrous flood and this particular kind of catastrophe has not occurred before in living memory, all of the relief actions that are undertaken by governmental agencies can be considered as special interventions. However, if flooding occurs quite frequently, or even reguarly, there will be officers working within agencies whose tasks in responding to the flood are prescribed and will have become routine. Thus, what is called an intervention relates to the frequency of occurrence of the event or sudden change. Whether a model handles the response to occasional and rare events by a programmed response or by manual intervention is actually not important. It is the scale of resources allocated, the speed of response required, and the degree of anticipation necessary that are important. These things can be investigated under either regime, that is by preprogrammed or manual intervention.

The exogenous variables corresponding to policy options should certainly be variables that political decision makers can actually select; if this is not so, the meaning of experiments in which variables are changed is unknowable. Governments can change levels of taxes, subsidies, and tariffs on imported goods. Governments can undertake capital investment projects that are not profitable in the short term. Governments can undertake training programmes whereby the distribution of skills in the populace is different from what it would otherwise have been. These are the kinds of policy variables that ought to be used in models.

It is generally possible to translate constraints of an institutional kind into mathematical relations governing particular variables. For example, bureaucratic delay in decision making can be modelled quite easily.

It appears that, to a very large extent, the behaviour of economic systems can be interpreted in terms of simple maximizing or satisfying of objectives for the individual actors, for example investors seeking adequate return on their investments, workers seeking adequate return for their labour, etc. It is a useful procedure to specify the model in these quite simple terms and then to make it more elaborate insofar as observed behaviour demands that more elaborate sets of goals are to be inferred.

In general, the interaction of the actors is through the various kinds of bargains that are struck in the economic sphere. This idea can be continued even into the non-monetary domain if necessary.

4. Structural aspects

a. Constructing the model

The same results could not have been obtained with a more highly aggregated model, because much of the output of a disaggregated model refers to specific individual regions that must be present in the model for the particular results to be obtained at all. It would be almost certainly possible to obtain the same results with a different structure. Omissions in the models are best decided pragmatically, that is by asking if the variable or relation is necessary for the operation of the model (and, if not, it should be omitted). However, it should be noted that some mechanisms lie dormant and are aroused only by the appearance of a specific stress. The fact that a given mechanism is unused in one run is no reason to discard it when it is known that it will operate in another. The type and extent of disaggregation is determined by the perceived problems, and the exploration of disaggregation is a learning process, that is one discovers new features of problems that lead to different experiments.

The readers of SARUM 76 must judge whether in fact the effort to compress and simplify the essence of SARUM in a form that allows for understanding rather than simply complex explanation has been successful.

SARUM 76 does not yield very much in the way of counter-intuitive results. Its value lies primarily in yielding the scales of effects — in other words, not so much answering the question of whether x will occur at all but how much of x. Additionally, the unexpected arousal of a causal link can provoke the modellers to reexamine their assumptions.

b. Consistency among regions

The same relations are used for all regions, but the parameters that are observed to apply in individual regions are used within the model, and thereby enable a match between the initial states of the regions and the initial states of the model to be achieved. The model is never precisely at equilibrium, but movement is always towards equilibrium, and the non-equilibrium behaviour is interesting in itself.

c. Adaptation of actors to changes

Most of the relations in the model can be interpreted as adaptation to changes in one or more of the variables. It is not a sensible question to ask how the model represents the fact that actors adapt to change, because they are adapting all of the time.

Changes in structural relations can be modelled by the inclusion within equations of terms that are not activated until particular features emerge in the external environment.

We are not yet sure of the best way of representing technological progress. Our investigations suggest that the proper representation of technological progress is probably the most important feature in the whole model structure.

There are no special strategems within modelling work that can help to answer the question of how to represent technological progress. This question will be answered by a more careful study of what happens in the real world.

d. *Model flexibility*

Experience suggests that it is most unwise to assume that a model can be successfully constructed as a one-time venture without any modifications arising from new insights and/or new policy questions. Therefore, the advice should always be to design in a modular and flexible fashion.

There can be no substitute in the modelling field for experience by individual modellers of the best ways to ensure modularity and flexibility. Some help is afforded by observing what others have achieved.

5. Testing the model

a. *Confidence in the model*

The parts of SARUM 76 that relate to physical constraints and physical variables of all kinds are the ones in which one can have the most confidence. These variables are also the ones that can be considered the most concrete. Subjective judgement enters in the postulation of relationships that are not immediately demonstrable from observed behaviour, but which are argued to be present because they are necessary for the general dynamic behaviour of the entire structure. These postulated relations are also the assumptions about which the modellers are least certain.

Provision for taking uncertainty of parameters and of data into consideration has been made through Monte Carlo runs of the model, so that the spread in the values of output variables can be assessed in terms of the histograms for values at given times in the future. Uncertainty in structure itself is an idea that has not yet been tackled. SARUM does indeed produce sensible results when subjected to noise, and, in the case of larger disturbances, the modelled behaviour bears good resemblance to real-world behaviour.

b. *Validation*

The operation of individual modules within the total structure can be checked for correspondence with reality. Thus,belief in the performance of the entire model depends essentially on the manner in which the modules are assembled. In some cases, subsets of modules can be checked for correspondence with real-world behaviour. Finally, acts of faith are required only with respect to model operation that carries one or more variables into regions never observed before at any time or place. This is not, in fact, very demanding. For example, the entire human race has, as far as we know, not suffered acute food shortage simultaneously; but individual groups of people have at all times suffered acute shortages, and even widespread deaths, from starvation. Thus, modelling a world in which food shortage is acute and universal is not novel in any fundamental sense.

c. *Calibration*

It is necessary in any modelling enterprise for correspondence with past behaviour to be demonstrated. This is only one of the separate tests required to give confidence in the validity of the model. It is of value primarily in eliminating structures that are incapable of providing the necessary correspondence with past behaviour. It is not sufficient of itself to provide confidence in validity.

d. *Sensitivity tests*

A part of the purpose of sensitivity tests is to allocate effort in refining the values used in the model in a more optimal fashion. Clearly it is better to allocate effort in refining the values that have significant effects on the output and to avoid allocating effort in refining the values of parameters that have very small or insignificant effects on the results. Some indications of preferred changes of structure also arise from results of this type.

The output of modelling work can be considered in terms of sets of sensitivity tests where the variations are applied to the exogenous variables.

6. Internal organization

Global modelling work is best organized in relatively small groups, but groups in which all the necessary disciplines are represented. It is doubtful if a group can operate as a single coherent modelling entity if its size is greater than about twenty.

Consistency between subgroups can be secured only by frequent interchange and discussion.

Global modelling groups should interact in the same way as the scientific teams in different research establishments that are investigating the same kinds of phenomena. Thus, there should be opportunities for publishing results, for conferences at which workers can meet and exchange ideas — and also opportunities for repetitions of each other's work. Cooperation beyond this level is neither necessary nor desirable.

All software that is subject to development must be maintained in such a fashion that the documentation is kept continuously up to date. The rules and procedures that have been devised over some years now by software houses provide the best examples to follow in this matter.

The general results of global modelling and the structures on which the results have been obtained should be published in scientific journals. Research institutions that contain both specialists in individual scientific fields and modellers can provide the best ground for interplay between the modelling community and the scientific community.

7. Relations between modeller and user

There is no general answer to the question of whether an explicit client should be identified. A client may be necessary in order to secure funds or to obtain approval for the work to be done. The modellers may believe that their work is only of value to particular decision makers, in which case it is clearly essential to work for specific clients. Conversely, it is clearly possible for the results of modelling to be of value to the world at large, in which case there is no special client and the customer can be regarded as mankind itself.

In the case of SARUM, approval had been provided for the development of a model, but without specific problems for which the answers were required.

All possible means should be used for communication with clients, that is, text, visual aids of all kinds, general presentations, private dialogue, and interactive exercises. This may involve the development of meta languages to assist understanding dynamic relations.

Ultimately the client has to take a part of any modelling exercise on trust. If a client were able to pass beyond this stage and not require any trust at all, he (or she) would have become a member of the team itself. The trust that clients should place in modellers' work ought to be derived from the modellers' track record. In modelling areas other than that of the globe, there is some evidence that this advice does apply. The most robust arrangement for advising clients derives from an open field in which modellers with different structures and starting from different viewpoints can offer their goods freely. Criticism is available from members of the peer group as for scientific work. Under these conditions there is far less risk of clients misplacing trust.

The findings of world modellers are likely to matter most with respect to changes in attitude. Thus people (as such) are likely to be the most important clients for world modelling. The group that is most interested in world modelling appears to be the senior staff of large companies.

The phenomenon of overselling is self-correcting. Overselling is followed by disillusion and a lack of receptivity to work that follows (even though the later work may be superior to what was oversold). This is unfortunate for the purveyors of work that is of high quality and is not being oversold.

The FUGI Model

The source of the response

The response dealing with the FUGI model was prepared by three people: Professor Yoichi Kaya of the Department of Electrical Engineering at the

University of Tokyo, Professor Akira Onishi of the Department of Economics at Soka University, Hachi-oji, Tokyo, and Professor Yutaka Suzuki of the Department of Electrical Engineering at Osaka University, Osaka, Japan.

Yoichi Kaya is the leader of the FUGI project and has the responsibility for its global input-output component. Akira Onishi has the responsibility for global macroeconomic modelling, which he has been studying for ten years. Yutaka Suzuki is in charge of the global resource model, which so far is concerned mainly with metal resources.

1. The purposes and goals of global modelling

The basic objective of global modeling is — as that of other regional modelling efforts — to envisage scenarios of the future, economy/resources/food/environment/society, as consistent consequences of the various options given a priori. The focus of each global model, however, differs from that of the others. The model complex FUGI is designed to focus attention on the future of the world economy and industry, especially in relation to the United Nations target of industrialization. In this sense, the model can be classified as neither predictive nor normative, but in between. In other words, the model can delineate the scenarios that are feasible in the sense that the foreseeable changes in any part of the society are taken fully into account.

2. Methodology

The appropriate methodology and model structure depend not only on the purpose of modelling but also on the availability of relevant data. The most pertinent model for the purpose of our work, FUGI, is a dynamic multisector model, which requires data on sectoral production functions that usually are unavailable in most developing countries. The authors eventually adopted a T-structured model complex, a macroeconomic dynamic model linked to a static input-output model. The macro framework of the input-output model (that is regional totals of final demand and/or exports and imports) is given by the econometric model. For the time being, the model employs neither qualitative variables nor physical constraints. It does not necessarily mean the authors underestimate the importance of these; rather, it is mainly due to the lack of reliable data.

3. Actors, policy variables

Our global model consists of two major integrated systems, a dynamic macro model and an input-output model. It provides information about possible future patterns of the world economy under various policy-oriented scenarios. For instance, in one scenario, decreases in the north-south gap and expansions of overseas development aid and private overseas investment coupled with modifications of industrial and trade structures, play significant roles as major policy variables.

How does the model represent the fact that actors adapt to changes? How should changes in structural relationships be considered? How should technological progress/technological change be represented? What other types of structural relationships are important in this context? Could 'structure sensitivity analysis' help to answer these questions?

The change in the pattern of parameters of input-output tables, reflecting technological progress, is predicted by using a modified regression analysis. The method being employed, the authors believe, is original, but it is still under development.

4. Structural aspects

a. Regarding past modelling efforts, could one have arrived at the same results (i) with a smaller model or (ii) with a different model (approach)? Have you developed a formalism by which to decide what to leave out of a model? Have you made any effort, after the model was constructed, to compress and simplify its essence in a form that would allow for understanding rather than simply complex explanation? What insights does the model yield that would not have been available through other means of analysis?

A few years ago we built a dynamic input-output model of normative type, making assumptions about the character of the sectoral production functions. The results are heavily dependent on these assumptions, which, as mentioned in section 2, are hard to prove valid. This is the main reason why we started to build FUGI, a new model.

As for publishing a simplified report that is easy to understand, we have not done it yet, but intend to in the near future.

b. How is consistency among regions guaranteed?

The model is a multiregion model in which interactions among regions are the trade matrix for the macroeconomic model. The exports and imports satisfy a world balance.

5. Testing the model

a. Which aspects of your model do you have much confidence in? Which aspects do you have less confidence in? Which aspects of your model require the most subjective judgement? Which aspects are the most concrete? What kinds of errors are to be expected (size and frequency)? What assumptions in your model are you least certain about?

What provision have you made to take the uncertainty of the parameters into consideration, the uncertainty in the data, and in the structural equations?

Has your model produced sensible results when subjected to noise and larger disturbances?

We believe that we have as much reliable data on the world economy to date as others have. For instance, we have collected several tens of input-output tables from various parts of the world, which are used to estimate future values of parameters. In this sense the model may be said to be heavily data oriented. Nevertheless, no one denies that there is still uncertainty on how the future will extend the past trends. It is certainly not a problem specific to the authors' model, but a general one common to all models being used in a predictive manner.

b. How can one be assured that the model will correspond to reality (validation)?

It is not our intention to make a forecast of the crystal-ball type, but to envisage consistent future scenarios corresponding to given assumptions.

c. How can one be assured that the model did correspond to reality (calibration)?

The model fits the past world economy well.

6. Internal organization

Global modelling requires enormous continuing efforts to collect data and a deep understanding of various factors in the world, both tangible and intangible. It needs generosity from its sponsors to understand the long-term character of the problem and the need for continuing efforts to revise the model.

The global modelling efforts are clearly of an international character, so cooperation among the few groups located in various countries, mainly in gathering data and exchanging information on crucial issues, is to be promoted.

7. Relations between modeller and user

It is worth noting that our global modelling efforts have been supported financially, not only by our own budget but also by other organizations such as NIRA (National Institute for Research Advancement) and UN ESCAP (the United Nations Economic and Social Commission for Asia and the Pacific). UN ESCAP is interested in using our model for the purpose of setting up regional targets with regard to the United Nations Third Development Decade (1980-1989).

Some other policy makers are also interested in our global model in that it might contribute to identifying internationally concerted actions to promote the growth of national economies, both at the regional and global levels.

The UN World Model

The source of the response

The response dealing with the United Nations world model was prepared by Anne P. Carter, Professor of Economics at Brandeis University, Waltham, Massachusetts, USA.

Anne Carter is one of the leading creators of the so-called United Nations model. In replying to our questionnaire, she stressed that her answers represent her own views and impressions, not necessarily those of her associates or of the sponsors of the work. To the extent that her answers are model-specific, they are based on her experience in building the United Nations world model at Brandeis.

After the United Nations model had been completed and delivered, the Brandeis Group began work on a second-generation model, called G2, that differs significantly from the first in structure and data requirements. While it is too early to give a full description of the experience gained in bringing up the new model, Anne Carter refers to it in her notes in answer to questions about future work.

1. The purposes and goals of global modelling

Since our understanding of the world is limited and our capacity finite, even 'global' models are necessarily specialized. A single model can deal only with selected aspects of reality. All models are 'open' with respect to some variables that are not explicitly endogenous to them. The UN model is open in that some variables must be specified exogenously before it can be computed. This is discussed further below. In addition there are countless unspecified variables — economic, political, social, physical, geological, and so on — that affect the state of the system.

A global model is not necessarily more complex or comprehensive and hence not necessarily more 'closed' than, say, a single-country model. Global models are all-encompassing only in the geographical dimension. The UN model covers all regions of the world in the sense that no place on earth is classified as being outside the scope of the model. It remains to be seen whether grouping the areas into the fifteen regions in the model is optimal for policy analysis. While each region block is specified in great detail, there are much more elaborate models of smaller geographical areas.

It is easier to fine-tune a model for a particular region than for the world at the same level of sectoral detail. Global modelling is appropriate for studying the interdependence of regions, the relations among activities in individual regions, and the changing context (say resource stocks, technology, etc.) that influences all regions.

In 1973 the United Nations asked us to build a global model as an aid in assessing their economic policy goals for developing regions. Their decision to sponsor the work was triggered by a grant from the Dutch government, which was concerned with the relations among global environmental and resource constraints and regional development prospects. Will the optimistic 'development decade' targets be infeasible because of dwindling resources and rising pollution? How will changing environmental and resource conditions alter costs of economic development and regional specialization? Should the goals be modified to take account of environmental dimensions that are important in themselves?

A global model seemed necessary, because the world economy had been showing growing evidence of interdependence. The model was conceived as a tool for conditional predictions and as an open model it was designed to explore the implications of exogenously specified policies or conditions for all the detailed variables of the system.

As is often the case in research, the definition of the problem evolved as the work progressed. At later stages, much greater emphasis was placed upon the interdependence of regional economic activities and much less on environmental variables. The shift was promoted by insights developed in the course of the research.

As the model took shape, we realized that we could take account of only a few of the many different types of environmental problems in the time allotted and that these few would not necessarily be the most serious ones. On the other hand, the model offered a potential for dealing with international trade, investment, and other economic policy issues in a global context, and we sensed that its prime contribution might lie in these areas. Needless to say, these areas have high priority at the United Nations.

In sum, the specific questions and the modelling effort to answer them were shaped by an iterative process.

It is misleading to consider a modeller's choice of methodology as a matter separate from his world view. They are both essential elements of his particular analytical 'style'. Similarly, users tend to find models reasonable tht do not contradict their own priorities and preconceptions. While this view of modelling denies the strict 'objectivity' of the activity, it acknowledges the personal element essential to scientific creativity.

Models do not make policy and they cannot supplant human judgement. They will be fruitful for policy analysis if they provide insight into the implications of explicit assumptions. The policy maker can use them as aids to his understanding. Under ideal conditions, a modeller or a user may select a given approach on the basis of his preconceptions, but later modify his preconceptions in the light of insights derived from work with the model. In reality, models are often misused to substantiate political or adversary positions. It is difficult to avoid the distortions that arise when models are used to 'prove' that a given political measure is good or bad without specifying the assumptions or conditions. The UN model has certainly been misused in this way. Misuse, in turn, tends to discredit the modelling effort. It is a matter of grave importance to seek ways to eliminate the abuses of models without eliminating the models themselves.

Few analysts, least of all global modellers, have the data they consider ideal for implementing their system. They have the choice of confining their efforts to pure theory or risking disapproval for 'uncritical' (note the wording of this questionnaire) use of inappropriate statistics. The UN model draws on cross-sectional statistical data to estimate longitudinal dynamic relationships. We have always been aware of major shortcomings in this approach.

In general, it is possible to improve the estimates of world model parameters both by intensive work with existing data and by gathering new data as time goes on. However, progress in this areas is expensive and requires long-term, serious, cumulative effort. Social scientists tend to place a high premium on ingenuity and a relatively low one on data work. Resources for building a new global model may not be plentiful, but funding for refining a data base or for cumulative refinement of an existing model is even scarcer. Despite the criticism of the UN model's crude data base, we have been asked to provide the same data to all too many would-be builders of new models.

The price system can be incorporated explicitly in a global economic model and the second-generation world model that we are building does have endogenous prices. Unfortunately, the price equations are subject to the same sorts of data deficiencies and analytical limitations as the equations of the 'real' system.

It is unrealistic to try to explain production levels without taking account of prices. It is also difficult to predict prices. The incorporation of price variables should not be viewed as a panacea for all the problems of economic modelling.

3. Actors, policy variables

The UN model is formulated in terms of regional economic activity levels and accounts, rather than in terms of individuals or policy agencies. The variables are not, in general, economic policy levers, but they are sufficiently detailed and concrete that it is usually possible to make a plausible bridge to economic policy. However, most regions are aggregates of several autonomous countries, and it would be unrealistic to expect uniform policies within a region.

In applications of the model, policies are simulated by assigning appropriate values to relevant exogenous variables and/or coefficients of the model. An increase in aid from developed to developing countries would have to be spelled out in terms of specific increases in the proportions of gross product contributed by individual developed regions and the shares of the aid pool received by individual developing regions. Migration to the cities could be simulated by increasing the value of the urban population variable.

Some policies are very easily cast in terms of the variables of the model, while others are more difficult to translate. To simulate the effects of changes in, say, tariff policies, it would be necessary to supply estimates of response elasticities or to derive estimates of direct quantitative impact from side models. The model would then be used to compute the general equilibrium effects of the specified initial changes.

The UN model has more variables than equations. For any given scenario or policy simulation, we choose which variables are to be treated as exogenous and which as endogenous. Values are then assigned to the exogenous variables and the simulation is run. The option of switching variables between the exogenous and endogenous categories gives significant flexibility for dealing with different types of policy questions.

4. Structural aspects

In bringing up the second-generation world model, we have dispensed with some detail in industrial sectoring and in specific resource accounts in order to streamline the model and to take advantage of data available only at a somewhat more aggregate level. On the other hand, we have added explicit price variables and a new set of variables representing income accounts. The net result is not a smaller model but rather one that is less elaborate in some respects and more elaborate in others. The decisions governing the reformulation were based on professional judgement rather than on any explicit formal criteria.

We have constructed a number of simplified versions of the model for didactic purposes. These take the form of schematic diagrams, illustrative equations, and graphs used in lectures. There is also a small model with hypothetical but plausible parameters that is used for testing and demonstration.

It is very difficult to give a comprehensive survey of the insights yielded by this model and impossible to say which conclusions might have been reached by alternative routes. The model is particularly well suited to keeping track of and tracing the implications of sets of simultaneous changes in interdependent regional economics. We found, for example, that energy-conserving equipment would not necessarily be useful in improving the balance-of-trade positions of developing regions, despite the fact that petroleum imports are a significant factor in their import bills. Since these regions also import a substantial proportion of their equipment needs, energy conservation would, at least initially, simply transfer their indebtedness from the petroleum to the equipment-exporting regions. Certainly this sort of insight does not necessarily require a formal model. The value of the model lies in its assistance to the human mind in keeping track of simultaneous relations, particularly where quantitative detail is important. In our model, consistency among regions is guaranteed because explicit equations representing interregional interdependence must be satisfied at all times.

The UN model does not model price substitution or technological change explicitly. However, the effects of these processes are studied outside of the model in the side models that generate changes in coefficients over time. Each of the four decadal systems (1970, 1980, 1990, 2000) has a different set of parameters representing the structures of national product accounts, input-output and labour coefficients, and import dependencies and export shares for all products. Changes in coefficients over time reflect expected changes in technology and anticipated responses to price changes. G2, our second-generation model, treats technical change as exogenous, as before, but allows for substituting imported for domestic goods in response to endogenously generated changes in regional, relative to world, prices of each traded product. It also allows for the endogenous substitution of labour and capital for energy as energy prices rise.

In the long run a modular structure is desirable, and may even be essential. It is important because it allows for cooperation among regional and industrial modelling specialists in a consistent framework. It is unrealistic to expect any single modelling group to incorporate the required expertise about every region of the world. In addition, policy groups in different regions are more likely to think in terms of a global model if they have been actively involved in shaping its development. We are aware, of course, of the political barriers to interagency and interregional cooperation, but continue to stress the urgency of this cooperation nonetheless.

5. Testing the model

6. Internal organization

7. Relations between modeller and user

Testing a global model involves backcasting, sensitivity analysis, and a running inventory of comparisons of projections with experience as they accumulate. There is much to be learned through study and comparisons of the results of global simulations with partial analyses done by specialized agencies. While there is general agreement that these activities are required, there seems to be no agency willing to finance the costly and continued work that is involved. Documentation is also very expensive and one may question whether a large investment in documentation is warranted without long-term commitment to developing the model.

Thus far, lacking adequate resources and continuing support, we have essentially built models to demonstrate what can be done. While it is now technically feasible to support complex large-scale systems, we have yet to build an institutional basis for making them effective.

5

The state of the art as seen by the artists: summaries of conference sessions and comments on the questionnaire

In theory, this chapter was to have been a summary of the questionnaire responses and the informal discussions at the Sixth Global Modelling Conference. In fact, it is a rambling discourse on the concerns, opinions, and interests of the modellers of large-scale social systems, closer in style to the writings of James Joyce than to an organized textbook on the practice of modelling. In spite of, or perhaps because of, its disorganization, it is also a rare and fascinating record of professionals in a new, uncharted, and difficult interdisciplinary field, discussing with unusual openness the way they do their work, their enthusiasms and uncertainties, their past mistakes and triumphs, and their hopes for the future.

The questionnaire and the conference sessions were divided into seven topics:

1. The purposes and goals of global modelling
2. Methodological problems
3. Actors and policy variables
4. Model structure
5. Model testing
6. Organization of modelling work
7. Modeller-client interactions

These topics seem distinct and separable, but they are in fact intertwined in the process of making a model and, as it turned out, in the conference discussions. The purpose of a model must obviously be related to its sponsor, and ideally also to the model's method, policy variables, and structure. Whether and how a model gets tested is influenced by such non-scientific concerns as funding, timing, and group organization. The conversations at the conference could not be confined to the scheduled topic of each session, and appropriately so.

Therefore an issue raised in one session often reappears and is resolved in a much later one. Topics come up in surprising places. For example, much of Session 3 on policy variables and actors is haunted by a deep methodological division among the discussants, which eventually becomes the centre of the conversation. Session 4 on structure includes a series of speculations about how to model technology.

Evaluators for each topic were assigned to compile the global modellers' questionnaire responses and to add their own comments about the state of the art in each area and possible directions for the future. In most cases the evaluators also chaired and reported the related conference sessions.

A word is necessary about the people providing the summaries and comments, many of whom have already been introduced in the preceding chapters. The global modelling community has grown to be a diverse and colourful group of people who know each other fairly well and know how to recognize and allow for each others' foibles. There is no way to find within that community perfectly unbiased rapporteurs and commentators on topics as divisive as the ones discussed at this conference. An 'insider' seeing a commentary by Dana Meadows or Michiel Keyzer or Sam Cole will know something about how she/he sees the world and how the report is likely to be slanted. For the roughly four billion persons who are not insiders, we as editors have provided small clues by identifying the authors of the various reports by institutional affiliation, country, and association, if any, with one of the global models. We can add that Onno Rademaker and Sam Cole have carried out extensive critiques of the World3 model, that Berrien Moore has set up and run several of the global models and prepared a review of them, that Stuart Bremer is just beginning a new global model in Berlin, that Bert Hickman is involved in linking national econometric models into a constantly updated global network, that Jennifer Robinson worked in the Meadows group but not on their global model, and that Paul Neurath had no connection whatever with any of the models or modellers until he came to the conference.

Beyond this, we can only warn the reader that the communications in this chapter are all biased, in different and interesting ways, and that the exercise of spotting the unconscious preconceptions and preferences of each author provides a direct way to achieve an understanding of the current state of global modelling.

Session 1. The Purposes and Goals of Global Modelling

Rapporteur and commentator: Donella H. Meadows, Resource Policy Center, Dartmouth College, USA (the World3 model)

This discussion is a summary of the views expressed by the global modellers in their responses to the IIASA survey, added to and elaborated upon by the participants at the Sixth Global Modeling Conference. The purposes and goals of modelling were discussed, not only in the first conference session, where this subject was scheduled, but also in virtually every other session, especially the one on the relation between modellers and clients. Therefore, I have drawn this summary from the proceedings of the entire conference.

The original questionnaire posed, essentially, four questions about the purposes and goals of global modelling:

1. What can be achieved with a global model? What are such models for? What can models *not* be expected to do?
2. What *specific* problems can and should be attacked with global models?
3. How can or should normative, descriptive, and goal-seeking aspects be incorporated into global models?
4. How should research funds and modelling resources be allocated in this field?

This summary takes up the answers to each of these questions in order. My personal comments are included in the sections labelled *comments* following each question.

1. What can be achieved with a global model?

> It is not clear what global models really are. They can be seen as mere ideological assertions, as mathematical exercises... as simple – however expensive – intellectual 'divertimenti.' But is it possible to see them as valid theoretical and methodological alternatives for the study of reality?
>
> G.A. Albizuri

The crucial question of what global models can do is a demand for a definition of the field and a clarification of the identities of its practitioners. It is equivalent to asking 'what are we all doing here?'. As might be expected, therefore, the modellers responding to the questionnaire and at the conference devoted considerable attention to this question. The fact that it was raised at all and debated with much energy indicates the new, unsettled, and unproved nature of the field. One cannot imagine similar conferences of other professional colleagues beginning with the question of what can be achieved with chemistry or economics or engineering.

The modellers mentioned a number of possible goals of global modelling. Each of the following goals was mentioned by at least one, and usually more than one, responding group.

a. *Scientific goals* (goals related to acquiring knowledge)

- Provide a clear, unambiguous, falsifiable scientific hypothesis.
- Identify inconsistent, sensitive, or missing pieces of information; guide analysts to important areas for further research.
- Encourage and train scientists in interdisciplinary research.
- Compile and process information of greater complexity than can be handled by the mind alone.
- Improve intuitive understanding of the behaviour of complex socioeconomic systems.

b. *Policy goals* (goals related to improving conditions in the real world)

- Provide conditional, imprecise predictions; general answers to if-then policy questions. (This goal was mentioned by virtually all respondents.)
- Communicate a world view or policy hypothesis.
- Explore the viability of alternative future systems.
- Discover unexpected problems from resource stress; eliminate physically impossible policy choices.
- Produce a management tool; select the best policy. (There was considerable disagreement about the feasibility of this goal.)
- Construct a neutral synthesis of many different ideas; provide a common language to express the assumptions of various actors in a complex system (for example labour, industry, environmentalists, consumers).
- Create an impressive-looking technical defence for a conclusion already reached. (Many modellers complained that clients and other modellers seek to use models for this purpose.)

Note that these goals apply to all forms of social system modelling, not global modelling alone. If global modelling has any special purposes separate from or beyond other kinds of modelling, they were more implicit than explicit in the discussion. I might postulate a few particular goals for global modelling that seemed to be assumed, if not stated, by the conference participants:

c. *Special goals for global models*

- Emphasize the interconnections among parts of the socioeconomic system or ecosystem that are normally seen as separate or independent; promote a global, as opposed to nation-state, world view.

- Encourage a longer time horizon in the discussion of social problems, and provide a means for envisioning long-term futures.
- Investigate problems that simply cannot be limited by the normal conceptual boundaries of academic disciplines or political nation-states (examples are problems concerning the oceans, the atmosphere, international communications, or international trade).
- Sketch a 'big picture', or qualitative overview, of future possibilities, which national or corporate planners can take into account in making their own more limited models.

Several modellers also listed goals that they believed could *not* be achieved by global models. They included:

d. Unachievable goals

- Unconditional, precise prediction. There was total agreement that social systems are not predictable in the sense that solar eclipses or missile trajectories are.
- Total comprehensiveness; inclusion of every possible relevant factor in complete detail. Models are necessarily simplifications. If they contained everything, they would be no easier to understand than the real world.
- Establishing goals either for individuals or for the total system. Goals may be represented in a model, but the model cannot tell what the goals should be.
- Objectivity. The statement of a problem and the choice of a method, as well as the specification of equations and decisions about what to omit, are influenced by the culture, training, knowledge, and personality of the modeller. (Some modellers claimed that one can *strive* for objectivity, at least insofar as not constructing the model to guarantee certain outcomes or to force certain policy conclusions. Others expressed amazement that anyone could believe in scientific objectivity and simply recommended that every modeller be as honest as possible about his or her biases.)
- Prediction of sharply discontinuous events, such as major earthquakes, thermonuclear war, visitors from outer space, or stunning new technologies, ideas, or charismatic leaders. At best, given the current state of understanding, a model might be able to trace out some of the implications of such an event, but not the exact timing, location, and strength of its occurrence.

Comments

When I first read these lists, I was struck by several inconsistencies and anomalies. For example, some of the possible goals of modelling seem to be mutually exclusive, or at least in partial conflict. Greater complexity in a model might interfere with clear communication. Production of a credible management tool might preclude creative formulation of new hypotheses. Comprehensiveness might lead to a model so complicated that it no longer can be understood intuitively.

An even greater anomaly appears in a comparison between the list of unachievable goals and the common expectations of the public, many policy clients, and even some modellers. Many people seem to be labouring under the assumption, or the hope, that computer models can be crystal balls that generate very comprehensive and precise pictures of what the future *will* be, and can tell us objectively and decisely what we *ought* to do. No modeller I know claims that his or her current models meet this expectation, and most deny that any model ever can meet it. Yet our public claims, private conversations, actions, and sometimes even our model designs often imply that we are striving for these unachievable goals and may be just about to achieve them.

I raised both these questions as topics for discussion at the conference. How often must we sacrifice one goal for our models to achieve another? And is it a problem that much of the general public and many policy clients expect from a model just the functions that are listed above as unachievable? I do not recall that any speaker dealt with these questions directly. But I felt them lurking under many of the overt disagreements among us. Should models be made from the top down or the bottom up? Are our models ready yet to be used? How can we deal with really complex models? Should we include sociopolitical variables for which we have no statistical data?

Our arguments on these procedural and methodological questions arise from our deep, unstated, assumptions about what we can know about the world, how we can be sure we know it, what a model is, and how a model is to be 'used'. I suspect the computer modellers of the world can be arrayed along a complete spectrum of beliefs about these issues. On the issue of model use, for instance, the range seems to extend from a vision of computer models consulted daily at the centres of power, directing detailed operations of social systems, to, at the other extreme, an assertion that computer models can never be more than conceptual toys that occasionally aid us in attaining a few useful general insights about how complex systems work.

2. What specific problems can and should be attacked by global modelling?

'I keep six honest serving-men
(They taught me all I knew);
Their names are What and Why and When
and How and Where and Who.'

said Rudyard Kipling, who forgot
the seventh serving-man: *With what?*
But let us first of all make room
for the most important one:
For Whom?

O. Rademaker

The respondents mentioned these subjects for future investigation:

(a) Managing human affairs during the transition period caused by the growing interdependence of societies and the finiteness of the globe.
(b) Exploring the band of potential trajectories the world system could take; identifying the resource constraints to these trajectories.
(c) Quality of life, especially how aggregate social, economic, and political processes are reflected at the level of the individual.
(d) How living conditions of the poor can be improved.
(e) How world population growth rate can be slowed, stopped, or perhaps even reversed (countered by a request for better models of the determinants of human fertility and for abandoning the antinatalist bias).
(f) Causes of maldistribution of economic output.
(g) The world arms race and its coupling with world trade.
(h) The transition from petroleum to renewable energy sources.
(i) The resilience of global ecological cycles to human impositions.
(j) The design of an economic system that does not depend on physical growth.
(k) Global communication patterns.
(l) Loan defaults and the international monetary system.
(m) The effect of multinational corporations on world trade and economic development.
(n) The linkage between the energy system and the economic system in modern societies.

Comments

I find it embarrassing to quote this list, because most of the entries are my own, and none, including my own, are specific enough to define a new modelling effort. Few of the global modelling groups in their written replies or the participants in the conference proposed any specific questions for future global models. I can imagine several reasons for this, which I will simply list here as hypotheses to be accepted, rejected, or added to by the reader.

1. The respondents may not believe that any more global models should be made.
2. We may be so preoccupied with our current modelling problems that we are uninterested in contemplating new ones.

3. We may be accustomed to receiving our modelling questions from policy makers, we may have little practice in posing our own problems, or we may feel it is not our professional role to do so.
4. We may be poorly trained in asking leading questions or in defining clear, feasible, limited modelling problems.
5. We may feel that it is philosophically unnecessary to pose specific, limited questions; we may believe that models can be fully comprehensive, or that one can model an entire system instead of a particular problem.
6. We may be unconnected with or uninformed about the real world.
7. We may not perceive any problems in the real world.
8. We may simply be unimaginative.

It would be interesting to compare the modellers' list of problems with similar ones drawn up by people in other fields, such as economics, ecology, politics, or law.

3. How can or should normative, descriptive, and goal-seeking aspects be incorporated into global models?

> If we were better at modeling, we would ask not how models should be done, but how the world works.
>
> A. Carter

This question brought forth a varied and sometimes heated set of responses. In part, the heat resulted from confusion about the words 'normative' and 'descriptive'. It was pointed out that all models are normative, in the sense that all are based, implicitly or explicitly, on the values and priorities of modellers and/or clients. All models are also descriptive in that they contain statements about the way social systems work or the way elements are interrelated. The real choice is between 'goal-seeking' models, which declare a specific societal goal and then use optimizing or other techniques to ask whether and how that goal can be achieved, and 'projection' or forecasting models, which simulate the system, assuming that it goes on operating as it has in the past or under specified policy changes. Goal-seeking models assume some target or objective function in the future and select policies that will lead to the target, whereas projective models assume policies and forecast what will happen if they are implemented.

The conference discussion suggests that there is room for, and probably need for, both types of models. Some modellers spoke strongly against the idea of a global objective function or even the use of optimization methods for long-term, large-scale social systems. Others found these methods useful. Some felt that models can be used legitimately to express a social vision or utopia; others worried about the credibility of such exercises.

There seemed to be some agreement that models can include (and some would say should include) representations of goal seeking *within* social systems. That is, one can represent the goals of the actors (individuals, organizations, nations), how these goals influence the actors' behaviour, and what happens when goals conflict or behaviours result in suboptimizing goals. However, there was considerable discussion of the problem of obtaining information on human motivations and goal-seeking behaviour, and clear disagreement on whether one should attempt to represent such behaviour if precise data about it are lacking. The absence of sociopolitical and behavioural elements in computer models was widely identified as a weakness in the field.

4. How should research funds and modelling resources be allocated?

> No matter how many resources one has, one can envision a complex enough model to render resources insufficient to the task.
>
> M. Mesarovic

This question was originally phrased: 'How would you proceed if you had (a) limited resources, (b) unlimited resources to spend on global modelling work?' Some interesting (but not mutually consistent) suggestions were given that may be useful for sponsors and managers of modelling projects to consider. For example:

(a) There is a full spectrum of modelling tasks, some of which can be handled by one person in a short time, others of which are much more demanding. *The resources should be determined by the task,* not vice versa.
(b) Resources could be conserved if modellers built more on each others' work and modified, or at least drew concepts and data, from already completed models.
(c) If possible, separate, independent groups with varying biases, methods, and goals should be given the same task, to allow a variety of approaches and viewpoints.
(d) Choose central team members with proven capacity for interdisciplinary work, rather than for depth in a given field. Depth can be added, as resources permit, by commissioning experts to do sectoral studies following a hierarchical order of priorities specified by the central team.
(e) Focus on areas where traditional disciplines meet (or fail to meet).
(f) Define the basic research problem carefully and specifically. Don't try to model everything.
(g) Concentrate on the central problem in the model, but include rudimentary representations of interlocking aspects — better a crude, shallow representation than a total omission.

(h) Allow for monitoring, testing, documenting, and updating the model; don't just make it and drop it.
(i) Have a first rough model running very quickly; then work on correcting and elaborating it.
(j) Reserve half the time and funding available for documentation.
(k) Give the model to the most competitive or antithetical modelling group for testing and evaluation.
(l) Always limit research funds, to force simplification and completion.
(m) Do 'bottom-up' modelling – start with subsectors and pieces you understand and build up from there, validating step by step, as a scientist would do. Then if your resources are limited, wherever you stop you will have a solid foundation of understanding.
(n) Do 'top-down' modelling – start with a rough sketch of the entire relevant system and then begin elaborating and adding detail, as a scientist would do. Then if your resources are limited, wherever you stop you will have a complete overview of the system.

Comments

I feel no need to add to or comment on this list, except to add to the top-down-bottom-up controversy the voice of a distinguished systems scientist who was not present at the conference:

> The more we are willing to abstract from the detail of a set of phenomena, the easier it becomes to simulate the phenomena. Moreover, we do not have to know, or guess at, all the internal structure of the system, but only that part of it that is crucial to the abstraction.
>
> It is fortunate that this is so, for if it were not, the top-down strategy that built the natural sciences over the past three centuries would have been infeasible. We knew a great deal about the gross physical and chemical behaviour of matter before we had a knowledge of molecules, a great deal about molecular chemistry before we had an atomic theory, and a great deal about atoms before we had any theory of elementary particles — if, indeed, we have such a theory today.
>
> This skyhook–skyscraper construction of science from the roof down to the yet unconstructed foundations was possible because the behaviour of the system at each level depended on only a very approximate, simplified, abstracted characterization of the system at the level next beneath.

Herbert A Simon
(*The Sciences of the Artificial,* pp. 16-17,
MIT Press, Cambridge, Mass., 1969)

Session 2. Methodological Problems

Rapporteur: P. Roberts, Systems Analysis Research Unit, Department of the Environment, UK (the SARU model)

Commentators: S. Cole, Science Policy Research Unit, University of Sussex, UK; O. Rademaker, Project Globale Dynamica, Technische Hogeschool Eindhoven, The Netherlands

This session directed attention to four separate questions, each concerned with a methodological aspect of global modelling:

1. How are global models to incorporate social and political aspects?
2. Should different methods be used for the short and the long terms?
3. How is the environmental dimension to be handled?
4. What new methods are emerging or ought to be explored?

1. Social and political aspects

Speakers were in no doubt of the importance of social and political elements; indeed, the general sentiment expressed was that of regret that modellers had so far failed to achieve satisfactory representation of these elements. Interest centred on the reasons for this failure and two strong propositions emerged. First, there is no encouragement from the client for modelling to be involved with live political issues, and this seems to be so irrespective of the nature of the client. Second, there is a strong peer-group reaction against qualitative elements appearing in models. Particular professional groups (economists were cited as exemplifying the attitude) deprecate approaches that are essentially non-quantitative. The response style is typified by a comment such as: 'If you cannot measure it, it does not matter', or 'If there are no data on it, leave it out.'

A more powerful form of this pressure appears in the idea that qualitative approaches are non-scientific and non-professional, so that whoever attempts to break away from the use of quantified variables puts himself beyond the pale and will be ignored by the scientific community.

In spite of the problems of quantification in tackling political issues, discussants were clear that action generating change takes place in the political arena, and therefore modellers concerned with change must address themselves to the political dimension.

2. The short versus the long term

There was general agreement that most short-term methods are inappropriate to global issues, but a stronger view came through clearly from several speakers. It was argued that some short-term approaches are positively misleading in that they distract attention from significant factors that need to be represented. For example, input-output with fixed coefficients can convey the impression of a smooth untroubled future, when it is the change of the coefficients that is of importance. The argument proceeded to the position that the nature of the problem perceived is conditioned by the method used to explore it.

3. The environmental dimension

The earlier part of the discussion on environment was concerned with a particular technique for handling pollutants within a general modelling framework; it consists of adding extra rows and columns to the conventional input-output table in order to show the additional unwanted outputs (sulphur dioxide, particulates, etc.) quantitatively and to exhibit waste disposal plus recycling activities explicitly. Although this method appears to offer the opportunity of bringing the environmental dimension within the purview of modelling, disenchantment with the method was expressed by those having actual experience of it. The point was made that one only inserts waste outputs that can actually be processed by known methods at acceptable costs. The method cannot represent the more intractable, pervasive, and troubling environmental insults that no one knows how to deal with.

From this position on extensions of input-output, criticism progressed to attack the idea of tackling the environment in 'bits': a bit of improvement here, a bit of abatement over there, etc. The climax of this attack was reached when a speaker identified the deep split between economic models, whose rationale is primarily exploitation of natural resources and, ipso facto, of the biosphere, and, on the other hand, ecological models whose principles are those of maintaining variety and dynamic equilibrium. The conclusion of this line of thought is that economic models must either be abandoned or at least drastically modified before a satisfactory treatment of the environment is possible.

One or two speakers who had direct experience of the administrators' problems in handling environmental pollution were concerned that the modellers should be able to contribute something practical and relevant to today's difficulties. They felt that abandoning the orthodox methods, unless there is an alternative that can be offered, is no answer. No resolution of the conflict between methods suitable for pragmatic decision makers and methods deriving from the ecological viewpoint was offered.

4. New methods

The last discussion on the potential for new methods contained a strong echo of the opening exchanges on social and political issues. Discussants were concerned, on the one hand, that methods capable of penetrating the less obvious variables become available, and, on the other hand, that do not, because of their esoteric character, exclude the social and political scientists. In spite of the difficulties that had earlier been noted of bringing in qualitative concepts, several modellers well known for their severely quantitative work spoke with encouragement about the need for qualitative modelling. For example, it was suggested that we did not actually need high precision for many of the issues and it would be enough to identify 'large' countries and 'small' countries rather than drown in a sea of numbers.

There was a marked lack ofenthusiasm for the more recent developments in mathematics. Catastrophe theory, developed by René Thom, although it may have the power to handle discontinuities (and the discontinuities in world history are the parts that interest us more than the smooth eras), was not well regarded by discussants.

(In his closing remarks, the chairman noted that discussion on each of the topics had to be curtailed because of insufficient time and that there were other important topics concerned with methodology which we had not reached. Clearly this is a most important aspect of global modelling and one about which the modellers themselves feel strongly.)

Comments (by S. Cole)

These remarks evaluate the responses to the questionnaire. I have had brief discussions with Peter Roberts and have taken up some of his suggestions, although he cannot be held responsible for any of the views I express. Professor Rademaker also sent a note concerning certain aspects of methodology and I have included his note in full at the end.

For reasons made clear immediately below, I do not attempt to 'evaluate' authors' responses on an individual basis, but try to draw out the central issues raised by their responses as a whole.

The first thing to say is that I have sympathy with modellers trying to answer disembodied questions on the topic of methodology. To my mind, questions of methodology cannot be treated in the abstract and neither, by implication, can their responses to a questionnaire on this subject. Choice of methodology is not without purpose, and this purpose is apparent from the responses to other parts of the questionnaire — for example on the issues to be confronted, the variables to be included, and so on. When each response to the questionnaire is taken as a whole, coherent and in some cases distinctive differences in modelling philosophy do appear.

A second matter to bear in mind is the terms of reference of an evaluation of methodology. To people outside the modelling 'world', all modellers are much of a muchness. To be sure, there may be different ideological underpinnings to their work, and as such it may be exploited by different interest groups, but these exploiters do not distinguish the finer points of methodology. Modellers, on the other hand, perhaps indulge themselves too much in these finer points. There are, for modellers, many fine distinctions to be drawn with respect, for example, to questions of programming language, machine (that is computer), fast algorithms for convergence, stability of solutions, etc. I do not propose to discuss these very real issues here, except incidentally. Presumably, however, even the advent of the messy-handwriting-reading, self-programming computer will not do away with the need for a very high level of technical expertise — and might even increase it. At the present stage of the game, modellers soon discover for their own purposes the relative advantages of, for example, different continuous simulation languages or man-machine interface devices, and, in any case, there is a lot of advice around.

My own frame of reference for evaluating a modelling methodology is that essentially it is a device in a process of communication, a heuristic device to help the individual researcher sort out his ideas, to improve interactions within diverse interdisciplinary groups who may be physically separated, and to communicate with an audience who, as noted above, may not appreciate some of the finer technical points. Thus, methods help to clarify arguments, help to build up a case for a point of view, etc. Incidentally, I take the position that, for most subjects considered by global modellers, there is no 'proof'. Models may help to lessen risks and ultimately provide better 'numbers' to use in detailed policy making. But first we should be sure that we have the *direction* of change likely to be induced by a particular set of policies correctly formulated. Again this comes down to a question of theoretical and ideological underpinnings. Some of the present models have to be wrong; they may *all* be.

The emphasis given to different aspects of methodology depends on the audience. A different emphasis would be expected from a model designed for a research experiment for fellow researchers (for example, a novel theory to treat unconventional data) compared with a study for the elusive policy maker who may view more sceptically (or cynically) such experimental work, in view of the fact that her or his political career rests on making the right evaluation of the implications of forecasts and projections. It is clear that methodology is an extension of the modeller — a model is in itself nothing without great mental effort in interpreting the results. The most successful forecasts (or so it appears from the track records of econometric forecasters) derive from modellers with a good nose for other trend indicators who are not afraid to let their better judgement outweigh the results of their more elaborate (but sometimes less sophisticated) models.

Even though separation of methodology and purpose poses many problems, the fact remains that it is possible to present contrasting views (and to different audiences) using a given modelling approach (or even a given model). This is a long way from saying that a 'best' methodology can be discovered in the abstract, without consideration of problem areas and particular theoretical formulations, available technical expertise, etc., etc. Here at IIASA, while we are surely focusing on tackling issues using computer simulation models, it has to be recognized that all of the global issues raised could be dealt with by non-computerized methods — probably leading to similar results. But even if only slightly 'better' results are obtained from our computer models, the marginal improvement to policies is valuable. A 1 per cent improvement in, for example, a forecast of the number of five to fifteen year olds in the year 1990 in any country alone, if it could be translated into money-saving terms, probably would pay off the research-and-development costs on global models to date. It is an open question whether such improvement is already achieved. We need to discuss, in the evaluation, the time within which these exercises are expected to bear fruit. I personally do not believe that we have yet made any quantum leaps with regard to actual quantitative forecasts. What I do believe is that global models eventually can improve the quality of discussion. Certainly global models are becoming part of the armoury in the ideological confrontrations within and between international institutions. Far from what some of us may have predicted five years ago, funding for global modelling exercises appears to be growing.

If we put these issues on one side, and yet still want to find common ground for discussion, there does seem to be a shared concern that emerges in the responses around the question of 'what is the correct level of detail in a model?'. And this question does get to the core of the debate about global modelling: heuristic tools or detailed planning aids? Related to this central question are others: what to put in and what to leave out of a model, the problems of data availability, and so on. Without wishing to gainsay the respondents to the questionnaire (as noted earlier I do not think it is reasonable to isolate methodological issues) the following answers to the central question may be characterized:

- work up to the limits of 'official' data availability,
- include *all* relevant variables, whether hard or soft,
- keep a model as simple as possible,
- include detail through a series of models and submodels at different levels,
- experiment and play it by ear.

These responses are not mutually exclusive, and the balance is a matter of judgement as the last opinion implies.

I propose to deal with each of the topics raised above in turn. Obviously an exhaustive commentary is not attempted, nor do I think is necessary for this audience. But the issues, I think, do reflect the concerns of the respondents.

- *Limits to data and classification*

I include classification along with data, because it is a fundamental part of theory creation. Many important insights have been derived because of new categorizations; many insights are blocked because of the difficulty of gaining data relevant to desired categorizations. Modellers who retain the SITC codes and variables are those working most immediately with (and funded by) government agencies. Thus, the idea of working detail up to the limits of official data is keenly sponsored in the Japanese study. Nevertheless, even the studies based on SITC are sufficiently different (for example the agriculture, manufacturing, and service sectors or regional groupings) for comparison of the results of different models not to be easily comparable.

To my mind, this raises competing problems: the need to have some standard for evaluating results and the need to get away from an obsession with data. Surely we have to have a different attitude in building long-term global models from that of statisticians engaged with short-term econometric models. We all know that the major uncertainties in models come from a few major variables: technical change, investment rates, possibly levels of trade, patterns of demand, and government activity *in the future*. Projecting these variables thirty to fifty years forward on the basis of twenty years (at most) of unsatisfactory (sometimes politically distorted) data gives only a poor guide for the future.

Constructing theories on the basis of data collected in the past is obviously essential, but what matters is how the interpretation for the future is set up. Crucial data relevant to important issues are still weak (for example income and wealth distribution data, output in the informal sector), but the situation is improving (for example through setting up of social accounting matrices for many developing countries). Global modellers can give a lead to data collection by demonstrating the possible importance of a variable using quasi-data.

- *Inclusion of all variables, hard or soft*

This issue goes to the heart of what is meant by a 'global' model as being not simply a set of regional econometric models but as something more akin to political economics or a sociological theory. To combine a set of econometric models as in Project LINK is reasonable for *short-term* projections, but not for the longer term. However, the further one departs from traditional economic variables in global models, the greater the cross-fire from orthodox (and influential) economists. It seems to me that modellers have drawn back from the more ambitious concepts of the early models, going too much for a 'quick sell' to policy makers, avoiding criticism, rather than confronting the real problems of creating new and relevant social theory.

From my perspective, most models deal poorly with social and political variables in particular. How important these are depends on the underlying theory. In some theories the economic variables appear to be almost independent of political events. But for most, there is a problem of how to integrate variables that are not easily quantified into the models. Should these be included explicitly or not? What is certainly necessary is to be able to show the implications of non-quantified variables for the quantified variables included in any model (and vice versa) and to show exogenous links. For example, in a given scenario, how may increasing job alienation affect capital productivity and what are the links between rates of technical change and trade? The second may be modelled formally; to do so for the first may be more questionable. There still remains the methodological problem of how to build up scenarios, how to iterate systematically between the formal model, the global picture, the desired image of the future, and the methodologically varied subprojects always associated with global modelling exercises.

- *Level of disaggregation*

Analytic tractability (that is producing a model that gives the essence of a theory in a simple algebraic form or programme) is only possible with high aggregation. This goes almost without saying, but again it seems to me that modellers have tended to go for high disaggregation without always first questioning whether the underlying theory and conclusions to be drawn from it make sense or even going back to check whether the central thesis is really substantiated by the model's results. As noted earlier, the first objective should be to check the plausibility of the story derived from the underlying assumptions of the model. Is the direction of change induced by one variable on another plausible, or have parameters simply been calibrated to fit in with prior expectations about the future? These questions may be raised with respect to all the central assumptions and policy variables of the models – directions of technical change, the economies of scale, the determinants of investment, the benefits or otherwise of trade and aid, the relation between increased global and regional per-capita income and domestic distribution or levels of absolute poverty. With models that are simply theories of capital accumulation and little else, this is not too difficult. As models become more complex (for example begin to take account of market and social feedbacks), results may require more than a second glance to be understood.

Incidentally, the question of calibration is, to my mind, side-stepped by all modellers, but, given my observations above about data, I will not dwell on this here, except to point out that there are no statistically satisfactory techniques for calibration and sensitivity analysis of models: the more detailed they become, the more difficult the 'fitting' of parameters, and the more implausible the procedures. As noted above, 'policy' modellers tend to seek a high level of disaggregation, because, for project evaluation or detailed sectoral policy, this is undoubtedly required. Nevertheless, much disaggregation appears to be spurious, since often groups of parameters are adjusted by blocks (of sectors) or blocs (of regions). This seems to negate much of the value of detail, although it is clear on practical grounds why this procedure is adopted. Not to do so requires the separate estimation of a very large number of parameters.

- *Models and submodels*

Attempts to resolve the problems of detail through a hierarchy of models and submodels are common. Whether a 'top-down' or 'bottom-up' approach is used may, however, affect the results. Simple recursive calculation of global or regional aggregates broken down by sector often gives surprisingly different results from systematically building up the global or regional aggregates from the sector or subsectoral levels. The relinking of these submodels depends on the nature of the problem, but in some cases the degree of connectedness among subsystems may be significant for the results. In other situations, this may not matter. For example, where small regions and small sectors make little difference to global aggregates, the global model may be used to provide an exogenous economic environment for the regional or sectoral model without introducing feedback effects. Indeed, this may be the best solution, since, wherever the output of the model is derived from the difference of large supply and demand aggregates, the errors may be high. This is often the case with the estimates of trade for developing regions in current models. Some disaggregation into submodels is needed, if only to divide up the work load in building a model.

While in general one must accept limitations of global models in confronting complex issues, certain characteristics of present models, in particular the higher levels of disaggregation now being attempted, in addition to creating inflexibility in relationships and demands for data that cannot be realized, may not add to the 'quality' of the results. For some of the global modelling studies, the level of disaggregation is high, but, even if the results were reliable, it would still not be possible to draw detailed policy conclusions at the level of existing administrative institutions. Given that this gap is likely to remain and a hierarchy of models to be required, all that is needed of the 'global' model is for the choice of aggregation to be such that the main quantitative results will not differ significantly if greater detail is added. More detailed phenomena may then be deduced from an enlightened interpretation of the results of the aggregated model.

That there are tradeoffs between the methodological opportunities and difficulties raised by the above set of issues is clear. In principle, one could tabulate these tradeoffs in the form of a matrix against the purpose and content of a modelling exercise; this might be useful to clarify the 'dispute' between advocates of different methods. When all factors are taken into account, it may well be that individual modellers have, on balance, sought out the most suitable method. The mutual 'criticism' which modellers engage in with respect to the suitability of each others' methodology is, in this sense, misplaced. This in itself does not imply that the models succeed in fulfilling their authors' intended purpose. The reason for this, however, in my opinion, would not be that the models fail on methodology, but rather on the treatment of crucial variables such as technological change, income distribution, and national and international social and political relationships. These, however, are issues to be explored by other evaluators. The problem with methodology is not so much one of 'methodology' as of 'approach'.

Comments (by O. Rademaker) on modelling ill-known social phenomena

The unavailability of knowledge — always a great handicap in model building — poses particularly compelling problems if certain social phenomena or interrelations are believed to be important, though little is known about them.

In Roberts' opinion: 'If relationships are in serious doubt, then it is better to avoid modelling the areas in which this ignorance is profound.' Instead of 'inserting explicit fertility mechanisms,... it is better to handle population increase in an exploratory fashion rather than to insert dubious causal mechanisms, though this does not preclude trying speculative mechanisms is an experimental style.'

Meadows eloquently discusses the dilemma posed by a model part that is believed to be important, but about which little is known. I share her belief that the fact that some social mechanism is difficult to model is no valid excuse for ignoring it altogether, because this amounts to introducing the assumption that the mechanism has no influence whatsoever. Almost any educated conjecture about such a mechanism seems to be better than ignoring it. However, this does not necessarily mean that the best procedure is to hide such a conjecture away in a computer model algorithm and to act as if the uncertainties no longer exist or have been reduced considerably. A lot of intuition, judgement, and plain guesses go into models of social systems. There is, in principle, nothing against this,

- *provided* it is unavoidable,
- *provided* one does not cover up that it is done, and
- *provided* all possible measures are taken to reduce the probability of undesirable consequences.

This again underlines that it is not only undesirable but even dangerous to use models as machines that are cranked to spew out masses of numbers and graphs, and that both *model analysis* and *dequantification** are indispensable as means for reducing the hazards of modelling social systems in the face of uncertainty.

Scolnik writes: 'Besides experts' judgements for defining plausible relations, it is important to deeply study the model sensitivity with regard to those variables and to present a spectrum of runs under different hypotheses, covering the range of their variation.' As I have tried to explain elsewhere, many new techniques have yet to be added to those developed in engineering and the physical sciences if the modelling of social systems is to mature.**

As one possibility I would suggest the use of flags to signal unusual conditions during simulations, for example circumstances that might activate mechanisms that are otherwise dormant and/or not easy to model with sufficient confidence. A very simple subroutine would monitor the relevant model variables and draw the attention of the user to unusual or critical conditions by typing out a message (*waving a flag*). During the model

---

*U64 (Rademaker): 'On the methodology of global modelling', Presented at the Fifth IIASA Global Modeling Conference.

**U67 (Rademaker): 'Modelling and simulation of macro-social systems for cybernetic purposes', Background paper prepared for the Opening Lecture of the Fourth World Congress of Cybernetics and Systems, Amsterdam, 21-25 August 1978.

U31 (Rademaker): 'Some notes on modelling and simulation of public systems', Paper presented at the same congress.

construction phase, one would not attempt to make any model of hard-to-quantify, unquantifiable or ill-known interrelations, but only to specify under what conditions these might come into play, thus specifying appropriate *flags* instead of dubious *causal interrelations*. During subsequent simulations, if, and only if, a flag is waved, the situation and the circumstances will have to be studied more closely in a variety of ways unrestricted by the paradigm of quantitative modelling.

Session 3. Actors and Policy Variables

Rapporteur: Donella H. Meadows, Resource Policy Center, Dartmouth College, USA (the World3 model)
Commentator: M. Keyser, Centre for World Food Studies, Free University of Amsterdam, The Netherlands (the MOIRA model)

The sesson chairman raised five questions for discussion:

1. Should actors in a social system be represented explicitly, particularly in their economic roles (consumers, producers, investors, etc.), or is it sufficient to represent the social phenomena *resulting* from these actions (price changes, capital growth rates, labour participation rates, etc.)?
2. Should the interactions among actors be represented recursively or simultaneously?
3. Is it necessary to define nation-states as principal actors? If one departs from the nation-state level, is there a sufficient statistical base to permit modelling?
4. Can policies, especially on the governmental level, be modelled endogenously, or is it preferable to use other techniques, such as scenario development or gaming?
5. Can the dichotomies revealed in the previous four questions be resolved, or can compromises be reached?

These questions, like all questions about how modelling 'should' be done, brought out strong differences among the conference participants. Some felt that representation of economic actors is essential, especially on the global level, in order to maintain necessary economic balances (such as monetary or material flows — all goods sold by one set of actors must be bought by another set of actors). Others felt that economic actors are irrelevant to problems of interest to them and that stock balances are more important than flow balances. Some felt that recursive interactions are more realistic and transparent, but this claim was countered by the objection that flow balances are difficult to maintain without simultaneous equations. Several people stated that representation of policy on the national level is essential. Doubt was expressed about this assertion, and the birth rate was cited as a vital social policy decision made quite independently from governments. Others suggested that the appropriate actors in a problem depend upon the problem, that actors shift their roles and functions over time, and that global modelling is done in order to *change* policy actors and their identities.

It was pointed out that these differences reflect extremely deep paradigmatic assumptions made by different modelling schools, assumptions so basic and unexamined that they are hardly expressible in words. The communication problem is compounded by the fact that key words, such as 'empirical', 'structural', 'policy', and even 'actor', have subtly different meanings for analysts using different methods to work on different problems for different clients. Then an interesting question arose: would modellers of different paradigms working on the *same* problem come out with the same result? There were various opinions here too, ranging from the expectation that the results would be the same to the assertion that they would be entirely different. Two conference participants reported on experiments to translate models of one paradigm into another, and both concluded that the attempts did not succeed because basic structure became too distorted in the process.

Other comments included a reminder of how many different actors on different levels there actually are in social systems, ranging from individuals to various levels of governments to multinational corporations and international organizations. To be stuck in any conventional set of categories, whether producers-consumers or national governments, is to be blind to many aspects of the real world. It was also pointed out that human actors process qualitative information, not quantitative, and therefore it may be more rational and accurate to represent them by non-numerical approaches.

Comments (by M. Keyser)

Question 3 relates to the way actors are represented in a global model. Table 14 summarizes the answers to questions 3 and 4(c) 1 given by the participating group.

According to the views of the Meadows and the Bariloche groups, which I will label empirically based, a global model should consist of equations describing phenomena, such as births, capital investment, or soil erosion. As far as human beings are involved in these phenomena, their behaviour is reflected in some equations, so that human beings ('actors') can be associated with these equations. According to this view, the association between equations and actors is, however, implicit and of secondary importance. In its second reply the Bariloche group has added to this that 'the notion of actor belongs to the theoretical framework of a model; some well established theories make an essential use of the notion, others can be formulated without explicit reference to it'. A global model however is 'a mathematical model or a computer program the theoretical background of which is not always made

Table 14. Summary of the answers to questions 3 and 4(c)1

Question 3	F/M	M/P	Bariloche
Who are the actors whose behaviour is endogenously modelled?	World consumers, producers, investors, governments, all reflected in rate equations	—	Four world regions
Who are the actors steering the model?	Behaviour of actors can be changed parametrically	Interactive user himself	Parametric changes in decision rules of regions
How are the actors represented? How does the model represent policy?	Through questions affecting the rates	'Cybernetic' representations	They can be associated with subsets of equations. Policy operates through regional constrained optimization
What policy options can the model test?	General options in the field of demographic, industrial, and agricultural policy; pollution control	—	Alternative courses of economic development, housing policies, educational policies
How do these policy options relate to the ones made by the leaders today? How many of policy variables correspond to policy alternatives that political decision makers can actually select given their power? How many are of an abstract synthetic nature?	Policies are relevant but are too general to be implemented in practice	—	Policies are related to current debate on development policies, but proposed policy change is quite drastic
How should political constraints of an institutional kind be tackled, methodologically?	Should not be disregarded but should not be taken as unchangeable either	Partly endogenous, partly exogenous	As constraints on goals
What do the structural equations depict in terms of the goals of actors?	The value systems as they are today	—	Socioeconomic goals
How do the actors interact?	Dynamic recursive interaction	—	—
Question 4(c)1			
How does the model represent the fact that actors adapt to changes?	Mainly exogenous but with some policy switches	Partly exogenous but with some policy switches	Endogenously through changed actions for reaching unchanged goals

MOIRA	SARU	FUGI
National consumer, agricultural producers, and governments	Regional workers, investors, resource allocators, governments	—
World government	Governmental agencies	National governments
Producer: constrained optimization; consumers, government, regression equations	Through reaction equations	Through equations of macroeconomic model
Alternative domestic food price policies. International food aid, supply and demand controls, price stabilization	Economic policy	Investments, aid, modifications in trade patterns
Policies represent a simplification but they are currently being discussed	They are actual policy options	
—	What cannot be endogenized has to stay exogenous	—
Government tries to keep income distribution at given level. Producer maximizes income, consumer just shows present behavior	Maximizing or satisfying level of net revenue	—
—	—	Through international trade
Endogenously. Too much structural change creates too many miracles.	Endogenous but problems with technical progress, policy switches	Parametric changes in input–output coefficients

explicit, and for most, if not all, of the known global models, that explication lacks necessary rigour'.

Quite opposed to this approach stands the one I will label axiomatically based. Here abstract actors form the basic units of the model (consumers, producers, trade union, etc.). Each actor pursues a (set of) targets and has at his disposal a set of instrument variables and a certain knowledge of the impact of these instruments on both his environment and his targets. One actor's behaviour is thus represented through one set of equations and another actor has another set of equations. In principle, one human being can be seen as different actors (a consumer or a producer), but this is rather undesirable. Optimization is a common tool. Once the behaviour of the different actors has been described, one has to interconnect them in some way. In a practical sense, this implies that one sees to it that each actor's input variables are generated somewhere in the model. In economics, prices are often the main input variable and a market-clearing procedure takes care of generating them.

The opinions expressed by the Bariloche group seem to be quite commonly perceived among the global modelling community. If one agrees with them, the direct conclusion must be that it is at this moment impossible to compare global models on theoretical grounds. One cannot compare them on empirical grounds either, because they generally do not have any predictive claims. One can therefore only classify them according to characteristics, like a botanist, and exclaim with the botanist 'let all the flowers flourish'. Obviously, as long as no common theoretical background is accepted, any proposition on how to build a global model remains with the domain of personal value judgements. Nevertheless, a frank exchange of such judgements is the only way to arrive at a common research programme.

The main advantage of the empirical approach lies in its flexibility. No abstract constructions are made; the only claim the model makes is that its variables are causally interrelated. The main weakness lies in the fact that there is no check on internal consistency. The approach operates on an equation-by-equation basis and the interrelations among equations obey no specific rules of rationality.

The axiomatic approach, on the other hand, puts emphasis on this rationality. A very large number of relations is usually embedded in one single optimization problem. From the empirical point of view, this seems a very artificial construction. Who is this aggregate optimizer? Can one speak adequately of, for example, revenue maximization at sectoral level, with only sectoral constraints? The axiomatic approach is less flexible, harder to implement numerically, and rests on very debatable assumptions, but it is explicit from a theoretical point of view. In my opinion, global modelling should in the future follow a more axiomatic approach (I mention three arguments for this, and of course I hope that, during the discussion, some counterarguments will come up):

1. Global modelling cannot be satisfied any longer with pointing out important phenomena and their interactions. It now has to pay

attention to policy advice, and policy advice has to be given to clearly identified actors in the real world. It is then practical if these actors can recognize themselves in the model in some way.
2. No one expects global models to come up with accurate forecasts. What the customer usually wants is to gain insight into the functioning of the world. This insight into the system prevails over empirical validity.
3. Most global models until now have come to the conclusion that the solution of global problems lies in the political sphere. It thus becomes relevant to identify political groups within the models, and for this the axiomatic approach seems better suited than the empirical one.

In my view, global modelling is a branch of social science. Global models should identify actors as basic units. If global models only describe phenomena, social interactions cannot be represented adequately and the power relations remain too implicit.

I believe that actors' behaviour should be represented endogenously. The social scientist should describe society. He may wish to learn from experiments with specific groups such as policy makers or consumers. Interactive computer simulation may be a tool in these experiments, but the final task is to represent the behaviour of the groups *within* a model.

I prefer to model the interactions among actors, not as a separate submodel but as the outcome of the integrated operation of the actors' submodels. Physical and financial balances can serve as integration principles. Steering the model should not only be effectuated through manipulating instrumental variables but also through introducing new institutional arrangements (cartels, alliances, common markets). What is called 'structural change' should be more than manipulation of the model's parameters. It should involve introducing new structural equations and generalizing existing ones.

All these statements are obviously very general and informal, but global modelling is a new field without any common axiomatic basis; therefore, a rigorous discussion would be misleading. What is needed (and obviously falls outside the scope of a general verbal evaluation) is a formulation of basic axioms in which the notion of 'actors' and 'interaction among actors' plays a central role.

Section 4. Structural Aspects

Rapporteur: Donella H. Meadows, Resource Policy Center, Dartmouth College, USA (the World3 model)
Commentators: O. Rademaker, Project Globale Dynamica, Technische Hogeschool Eindhoven, The Netherlands; B. Moore III, University of New Hampshire, USA.

The chairman introduced such topics as modular construction and regional disaggregation, but the discussion centered almost entirely on the question of how technical change can be represented in mathematical models. Among the comments made were these:

- We usually make technology exogenous in our models, primarily because we view it as a policy instrument and because we have no theory for making it endogenous. Can we do better than that? If technology is endogenous, is it a function of research expenditures? Of education?
- Technology probably is dynamically very different in the agricultural and industrial sectors. In agriculture, research is largely government sponsored, and the increasing stock of public knowledge has to be transferred to a large number of small users. In industry, advances in knowledge are achieved privately, licensed and kept semi-secret, and applied almost immediately by a few large users.
- Aggregate representations of technology in economic models are misleading. The Cobb-Douglas function is entirely uninformative and unrealistic. Technology affects physical relations, but economists see it only through its indirect effects on monetary relations. The connections between the physical and the monetary changes are simply not understood, not to mention the relations between technical change and social characteristics such as alienation, unionism, and goal satisfaction.
- There are two distinct problems in representing technology. First, how do new technologies originate and, second, how do they spread to affect society? (And one might add a third. What are their unforeseen side effects?) The problem with representing and linking these is that the first is a specific, highly differentiated microlevel process and the second occurs on the macrolevel. Relating processes that occur on different hierarchical levels has always been one of the most difficult problems in any science, from physics to economics to global modelling.
- It is generally assumed that technical change is *the* crucial variable in models of growth and development. This assumption, and also the way in which we represent technical change, is not scientific: it is a pure act of faith. To the extent that we include acts of faith in our models, we urgently need some methods for mapping and comparing and communicating our assumptions. Otherwise, our acts of faith are likely to be accepted unquestioningly as solution strategies.

Comments (by O. Rademaker) on question 4(a)

The questions and answers make up a more coherent pattern if the questions are reformulated in this way:

1. Why and how was it decided to leave what out of your model(s)?
2. Would a simpler model have been possible a priori?
3. What about model simplification a posteriori?
4. How was the quintessence of the whole work extracted and communicated?
5. What (new) insights were obtained?
6. What other model/approach(es) would have yielded similar results?

- *About model simplicity*

Some people love simple models. Others love complicated models. From a professional point of view, both kinds are very likely to be wrong. One has to be exceptionally gifted — and preferably very lucky — to derive great results from a simple model, and maybe one also has to be very talented — and lucky — to remain master of a complicated modelling exercise and extract the best results.

From a professional point of view, the following aspects merit consideration:

- the requirements following from the problem formulation;
- the prospects of different modelling methods;
- the available knowledge about the system, (that is about the existence/availability or non-existence/non-availability of: variables (problems of definition!), interrelations between variables, data quantifying variables and interrelations);
- the knowledge required for model building: the same subdivision as the previous item;
- the requirements following from the results of the model studies and, in certain cases, from comparisons with reality;
- it must be recognized that any model-based study is bound to be coloured by the views of those involved, as Meadows argues so convincingly in her response.

Hence, things may be omitted from a model for very different reasons, for example things that are:

(a) not required by the problem formulation,
(b) incompatible with the method used,
(c) hard to include for lack of knowledge,
(d) superfluous with respect to the model in question,
(e) not required in connection with the use of the results,
(f) not compatible with the views of the modeller(s) and/or their clients.

Likewise, things may (have to) be included in a model for more or less opposite reasons.

Note that it makes some sense to distinguish between the *a priori* and the *a posteriori* situation. In the initial, *a priori* phase, the model's foundations have to be constructed without knowledge about how it is going to behave and be used, what kind of answers it will have to provide to what kind of questions. Hence, the considerations mentioned under (c) above do not come into play. In later phases, when the model is running, these considerations can and should influence the development and exploitation of the model, which makes the *a posteriori* situation fundamentally different. In my opinion, this distinction furnishes a strong motive for getting a model running as soon as possible — as is also advocated by Meadows — however shaky its basis may be and in spite of the risk that inexperienced and/or uncritical practitioners may mislead themselves and others.

● *1. Why and how was it decided to leave what out of your model(s)?*

Roberts refer to point (a) when he writes with reference to SARUM: 'The type and extent of disaggregation is determined by the perceived problems', and adds: 'The same results could not have been obtained with a more highly aggregated model, because much of the output of a disaggregated model refers to specific individual regions which must be present in the model for the particular results to be obtained at all.' In section 6 below we shall reconsider the important interrelations between problem formulation and modelling approach.

It is remarkable that hardly any explicit reference was made to point (b) things that had to be left out because of the modelling methods employed. The interactions between problem, method, and model are still largely 'terrae incognitae'.

On the other hand, the (un)availability of knowledge affecting the model, point (c), is mentioned in one way or another by most groups. It is clear that the logistic problems are great. They may become particularly compelling if certain parts of a model *have* to be constructed, as mentioned under point (d), and probably discussed extensively in connection with question 2.

Roberts writes: 'Omissions in models are best decided pragmatically, i.e. by asking if the variable or relation is necessary for the operation of the model (and if not it should be omitted)'. Possibly, Scolnik *et al.* are thinking along the same general lines when they write: 'Basically, the criteria for deciding what to leave out of a model are more normative than formal', adding 'however, if from theoretical considerations one intends to explain a certain variable as a function of others, formal criteria exist for eliminating independent variables which do not bring additional explanatory power', which is, in actual fact, the only response concerning a *formalism* for omitting something from a model.

Meadows writes: 'Our primary techniques for keeping models simple are probably training and peer pressure', adding, 'Students are forced to limit their models to only one or two levels or feedback loops', which makes one wonder how those students are later taught to unlearn such a simplistic notion. The first of her 'formal methods for attaining simplicity' is: 'Careful and deliberate definition of the problem', which is of very great importance indeed. The other formal methods mainly deal with later stages in which a model is already operational.

Finally, things are included in, or excluded from, a model because of the views of the modellers. As Meadows writes:

> Human beings are unable to think without... conceptual constraints. All problem-solving efforts, whether mental or formally mathematical, are limited by the set of *a priori* expectations about the world that has been labeled world view, paradigm, or Gestalt. There is no way to transcend this limitation, but its effects can be minimised by a careful matching of the real-world problem with whatever modelling technique is most closely suited' and 'I would say that the single worst problem in the field of modeling at present is the inexperience of modelers in examining their own assumptions, realizing their lack of objectivity, and understanding the relative strengths and weaknesses of their methods.

I think every modeller ought to have these words engraved in her or his mind.

● 2. *Would a simpler model have been possible a priori?*

In general, the answer (if given explicitly) is: no, not for the purpose we had in mind.

Meadows writes (abbreviated by me):

> Forrester's World2 model leads to very similar conclusions and is considerably smaller than World3.* *After* making World3 we learned enough about its essential mechanisms to make a very condensed model that exhibits the same basic behaviour. We could not have been able to produce such a condensation without constructing the larger model first, of course, nor would we have had much confidence in such a crude model. Nor is it very useful for elucidating different kinds of policies.

The 'methodological bias favoring simple, transparent models' obviously requires delicate restraint.

● 3. *What about model simplification a posteriori?*

In one way or another, most respondents have simplified or modified their model(s) in view of the results they were producing. As described by Meadows:

*This ignores the fact that the internal working of the models is different; for example the rise and decline of the population in the standard run is caused directly by the economic level (capital investment) in World2 and largely through the agricultural sector in World3.

Our formal methods for attaining simplicity include:

— Careful and deliberate definition of the problem, including a sketch of the problematic or expected behaviour of the main variables over time (reference mode). Inclusion or omission of a variable is then always related to whether it could possibly influence the reference mode.

— Initial production of a crude comprehensive running model that approximates the reference mode. We aim to have this model running within a month or so of the beginning of any project. Ideally this first model will contain the minimum causal structure necessary to produce reasonable behaviour. It is then elaborated, disaggregated, and expanded until we begin to have confidence in its structure.

— For difficult cases we have to use trial-and-error procedures (for example, population age disaggregation in the World3 model).

After remarking that 'omissions in models are best decided pragmatically, i.e. by asking if the variable or relation is necessary for the operation of the model', Roberts adds an important warning: 'However, it should be noted that some mechanisms lie dormant and are aroused only the appearance of specific stress. The fact that a given mechanism is unused in one run is no reason to discard it when it is known that it will operate in another.' In his response to question 5 (d), Roberts touches upon the link between model modification and sensitivity analysis.

Not much else was answered in connection with model simplification *a posteriori*.

● 4. *How was the quintessence of the whole work extracted and communicated?*

'By writing the book(s) we have written' seems to be the general answer.

With a view to the future development of social systems modelling, I want to stress one important point in connection with communication: whenever modelling results are being published, the model and its data base *must* be specified clearly and completely at the same time, so as to enable the scientific community to inspect the model and to test the results. Forrester set a shining example, which, unfortunately, was followed by few. A year or two after the publication of *The Limits to Growth*, we still knew only one thing for certain about World3, viz., that, when finally published, it would be different. And also most of the other efforts came nowhere near the standard set by Forrester (common in engineering); I feel sure that, in spite of the presentations at IIASA, none in the audience were able to reproduce the Mesarovic/Pestel model (which did not even appear to exist yet as a single computer model at that time), nor any of the other models, except, perhaps, SARUM and MOIRA. On several occasions, important questions could not be resolved during an IIASA symposium, either because no programme listing was available or because it did not give enough clear information.

Modellers must realize that a mere programme listing does not satisfy the minimum requirements of documented communication. Future modellers should do a much better job. Perhaps the most important practical lesson to keep in mind is that adequate, up-to-date documentation requires an enormous effort during any project. As Meadows remarks in her answer to question 1: 'If I had limited modelling resources' one 'half would be reserved for documentation of the model'.

● 5. *What (new) insights were obtained?*

Here are selected points:
Linnemann: 'The model is a learning tool for the modeller'. It helps to select among 'several intuitively acceptable options, it confronts the ideas with actual data and because of its computerization, it makes it possible to find out conditions under which the conclusions are false'. He adds, a little sadly, that modelling requires so much effort 'that not much time is left for "real" thinking'.

I am inclined to add that, in general, the same is true for thinking about *alternatives* to the modelling *approach* that was chosen (see below).

Roberts mentions 'the unexpected arousal of a causal link' and stresses that the value of SARUM 76 'lies primarily in yielding the scale of effects, rather than counterintuitive results'.

Scolnik makes a similar remark: 'But what the Bariloche model gave were the *trajectories* leading to... goals. It also *quantitatively* evaluates the complex nonlinear feed-backs...' while fulfilling 'the task of proving that the barriers for achieving the satisfaction of the basic needs at world-wide scale are more socio-political than physical'.

Meadows writes: 'Many other people have arrived at the basic conclusions of World3 via mental models, although their expression of these conclusions was necessarily less precise and systematic.'

● 6. *What other model/approaches would have yielded similar results?*

Many a human goal-oriented activity seems to be characterized by a mutual convergence of goals and achievements: while the latter are rising — often tantalisingly slowly — the former tend to be lowering, often almost imperceptibly, and when achievements and goals have approached each other 'close enough' the result is 'satisfactory'. Mental fatigue and saturation may play an important role in this convergence process; others who have not undergone this process may not be nearly as satisfied. This is well known in education, research, and many other activities.

At the other end of the scale, dissatisfaction with one's achievements may provide equally inconclusive evidence. Hence it remains to be seen whether much can be learnt by posing the question to which this section is devoted.

After a few remarks abut 'doomsday models', Scolnik *et al.* state categorically that 'It would not have been possible to use another approach' to prove that the barriers against the satisfaction of basic needs are more sociopolitical than physical and that the goals can be reached under certain conditions, because this task requires an optimizing approach.

Like Meadows and Scolnik, Roberts and Linnemann also acknowledge that more or less similar conclusions might have been obtained by intuitive reasoning or models having a different structure; they do not compare their approach with other possible approaches.

It is interesting to note that, even in the young field of world modelling, the choice of the approach is often influenced quite clearly by the force of habit or tradition. Like the other models, the FUGI model also reflects this: the principal motivation for the model recently proposed obviously is that much groundwork had been done to build earlier models and that these were obviously in need of coupling and further improvement. And the earlier models were firmly rooted in established economic modelling techniques.

To sum up: the question that was asked might have generated a lot of insight, but it did not. The field is still characterized by an embarrassing variety of possibilities: a great variety of conclusions, some rather contradictory, each of which can be reached by a great variety of approaches. It has been said that certain models may reproduce any person's signature, provided a few coefficients and functions are chosen judiciously and suitable variables are plotted against each other.

Yet, the various approaches and conclusions do have certain traits in common. For example, no model indicates that a decrease in the world's population is likely to occur within half a century. In engineering, model-based studies often help a great deal by indicating what is *not* likely to happen. Maybe the same is true for similar studies of social systems.

Another trait the various approaches have in common is the interrelation between the problem formulation and the approach used. The description by Meadows is well worth pondering:

> As in virtually all other modelling efforts, the method came first and shaped the question we asked, rather than vice versa. We were practicing system dynamicists, attuned to observing long-term basic dynamic trends in social systems, . . . The purpose of our model . . . was exactly matched to this method, of course. If we had not been system dynamicists, we would not have asked such a question.

• *The seller's attitude and the buyer's attitude*

A seller's attitude would be: 'I have this thing (a component, a method, a programme, a theory); for what purposes can I use it?' A buyer's attitude would be: 'I have this goal (for example the alleviation or elimination of a problem manifesting itself in practice); what can I use best to solve it?' Of course, everyone is partly seller and partly buyer, but in most people one of the roles seems to predominate most of the time.

My personal conviction is that most applied scientists ought to be *buyers* most of the time. This implies they should be intimately familiar with all the important methods and – rather than forcing every problem into the deforming straitjacket of a single method they happen to know best (I call this *the Procrustes-bed approach;* it may also be characterized as *falsifying a problem in order to make a preselected method beget a solution)* – they should take an eclectic attitude with respect to the flora of available methods.

Comments (by B. Moore, III*) on question 4(b). How is consistency among regions guaranteed?

This is a rather awkward set of comments to write. It was to have been my comments on the respondents' answers to the question: 'How is consistency among regions to be guaranteed?'

However, the replies have left me somewhat perplexed:

1. 'Regions are not disaggregated in World3' – World3.
2. 'The consistency among regions is guaranteed by appropriate representation of interregional relationships' — Mesarovic/Pestel.
3. 'The model is a multiregional model in which interactions among regions are the trade matrix for the macroeconomic model and the exports and imports satisfying world balance' – FUGI.
4. 'In MOIRA supply and demand are balanced. The balancing mechanism is world market price for food. MOIRA is a multination model and the world market for food links the nations to each other' – MOIRA.
5. 'The same relationships are used for all regions but the parameters which we observed to apply in individual regions are used within the model and thereby enable a match between the initial states of the regions and the initial states of the model to be achieved. The model is never precisely at equilibrium but movement is always towards equilibrium and the non-equilibrium behaviour is interesting in itself' – SARUM 76.
6. Here is the response with respect to the Bariloche model (emphasis added):

   We have used, like in the Leontief model, the pool approach, which has inconveniences (like the other known approaches). However, an emphasis on autarchic development gives some additional support to this approach.

---

*These thoughts reflect, in part, some of the insights that were gathered both individually and collectively by the Pacific Basin's group at the East-West Center in Honolulu with which I was associated from January to June 1978. Members of that group are: Dr. Ada B. Demb, Dr. Hisashi Ishitani, Dr. S. Langi Kavaliku, Mr. Budi Sudarsono, Dr. George Anthony Vignaux, and Mr. Nicholas R. Miller. However, because we have not been together since June 1978, they were not consulted about this paper.

> The consistency is not guaranteed, although it is not very relevant due to the assumption that the net balance of payments goes toward equilibrium (it does not mean zero trade). In fact, it is trivial to devise methods for obtaining consistency, but all the known ones are arbitrary and not based upon a theoretical framework.
>
> Without any doubts, *it is one of the weakest areas in global modelling.* Due to this reason, the structure and mechanisms of the international system are a research priority for our group, in order to be able to improve the next generation of models.

Fortunately, in some sense, this latter reply (and Professor Mesarovic's remark that 'One of the principal challenges in global modelling is how to represent the interrelationships [among] different parts of the world and different issues') gives me some manoeuvering room to sketch some of my thoughts about a slightly broader question. Namely, how should consistency and/or reasonableness *both among and within* regions be achieved? And, more generally, how should we choose regional disaggregations in the future?

These questions strike me as being non-trivial in complexity and important with respect to global modelling generally.

Purposes

The vastly disparate regional schemes used in World3 (1 region) and MOIRA (106 regions or nations) are largely a result of the modellers' different purposes. World3 was used to assist in examining which types of physical-growth modes in global population and materials output lead to stable, rather than unstable, behaviour of essentially global indicators, whereas MOIRA was designed to facilitate an understanding of the problems associated with the '(d)evelopment of food deficits in poor countries in the next four decades'. It can also be used to examine '(t)o what extent and how do the agricultural policies of the rich countries affect the development of food production and food consumption in the Third World'. The dissimilarity in purpose alone appears to be sufficient to yield the markedly different regional schemes. Consequently, it does not make sense to criticize a regional scheme out of the context of the modellers'/models' purposes.

Methodology

Though it may be less apparent, the choice of methodology influences the type of regional scheme. The United Nations/Leontief world model and the FUGI model seem to me to be near the limits of regional disaggregations (fifteen regions) if one is going to employ regional input-output tables that have at least moderate sectoral detail. I would find the application of this methodology at the level of regional disaggregation in MOIRA difficult to digest.

Actors, policy variables, and users

The Mesarovic/Pestel model exhibits clearly the influence of the choice of actors, policy variables, and users on the formulation of regional schemes. First, at the outset, Mesarovic and Pestel had as their central theme the usefulness 'in practical policy analysis' of their model(s). This desire for utility was paramount in pursuing different users. Further, their unique approach to regional disaggregation, which allows the subregionalization of one of the existing twelve regions, is an interesting recognition of both the curse of dimensionality and the terms of today's political realities, under which most of the key decision makers and policy-leverage points operate at the national level and in certain cases at specific regional groupings. (Though this may not be precisely the correct place, I should call attention to Professor P.N. Rastogi's article 'Societal development as the quintessence of world development' in *Problems of World Modeling – Political and Social Implications* (Eds. Karl W. Deutsch, Bruno Fritsch, Helio Jaguaribe, and Andrei S. Markovits), Ballinger, Cambridge, Mass., 1977. The article seeks to establish that 'national societies are the logical units of analysis in a study of the world system'.) The application of the M/P model to science policy questions for the Federal Republic of Germany is a perfect example of this technique. (See Chapter 3 of *Concepts and Tools of Computer-Assisted Policy Analysis* (Ed. H. Bossel), Brikhauser, Basel, 1977. See also the final report of the project 'Anpassung des Mesarovic-Pestel Weltmodells zur Anwendung auf Fuschungs- und technologierelevante Fragestellungen aus der Sicht der Bundesrepublik Deutschland', Institut fur Mechanik, Technische Universitat, Hanover, 1976.)

Some of the difficulties

It appears, then, that regional decomposition is interwoven with the primary conceptual issues that must be faced in global modelling. No single scheme stands out as an ideal; such would be non sensical, for regionalization is tied to model purpose, and there is no one ideal purpose.

Each system does have certain key difficulties, and there are some problems that are shared by all. Using only one region is likely to result in the greatest smoothing of the data, with a loss of precise short- to intermediate-term behaviour, and hence with a corresponding validation difficulty over shorter time horizons (up to thirty years). This difficulty is important, in part, since there is an apparent unwillingness for most persons to think over extended time horizons.

At the other end of the scale, where the number of regions begins to approximate the number of nations, lies the curse of dimensionality, both in a purely computing sense as well as the general usability of the model. The sheer numbers alone will be particularly burdensome if the model is going to disaggregate regional populations, the economies, utilizations, trade of resources, etc. The detailed trade algorithms in several commodities for national-level disaggregated global models alone will likely be either of such simplicity as to be out of keeping with detailed regionalization or so complex as to be unstable. For example, would one include for region-specific detail over one hundred different demand (or income) elasticities? Tuning for stability will likely have to be so robust that physical and social meaning will be left behind; furthermore, there will probably be several near-stable points, and choosing among them, assuming one can find them, will be difficult.

Since national accounting and its reporting varies greatly, what does one do — and this is true for all schemes — about the widely disparate sets of data across the set of nations? What is the appropriate reconciliation between the levels of information in presently available data sets for Denmark and China? (Professor Richard Stone provides an interesting discussion of this in his paper 'Major accounting problems for a world model', which forms Chapter 7 in *Problems of World Modeling - Political and Social Implications* (Eds. Karl W. Deutsch, Bruno Fritsch, Helio Jaguaribe, and Andrei S. Markovits), Ballinger, Cambridge, Mass., 1977.) Is it reasonable for each region to be given the same basic structure (that is, for example, a seven-sector economy, with capital and capital-output ratios by sector)? How should capital be treated? Exchange rates and depreciation? As disaggregation extends, these questions, depending upon one's sense of humour, become more difficult to avoid, more difficult to average away, or more difficult to 'sweep under the rug'.

Finally, there is the general difficulty of measuring the sensitivity of a model to different regional schemes. For the models at the extreme (World3 and MOIRA) of regional disaggregation, this difficulty may be vacuous, since the purposes of the model force a unique choice for regionalization but the question of this uncertainty is important for modellers who choose moderate region-disaggregation patterns (say, ten regions). For me this is a particularly disturbing point, since I have not found a clearly stated and formally reasoned basis for particular patterns.

Some vague suggestions for the future

For the moment, I shall assume that the purposes for which we shall be constructing global models in the near future will necessitate at least the moderate degree of regionalization exhibited in the first M/P model. In part, this assumption is supported at least implicitly, by all of the respondents. The Bariloche model has run under the Leontief fifteen-member regionalization scheme (see *Problems of World Modeling - Political and Social Implications,* p. 39); Professor Meadows' lucid response to the question 'What are the most important problems accessible to the field that have yet to be analyzed by global models?' seems to imply the necessity for regionalizing global models, and SARUM has appeared with strata two and three subdivided, so that the resulting regionalization was not unlike Mesarovic and Pestel's.

The issue of regional structures in global models is and will continue to be with us; therefore, because of its inherent problems – and without wishing to imply any specific criticism of the various regional schemes – I would like to suggest that the entire issue needs to be reconsidered. In hopes of sparking this reconsideration, let me sketch a possible structure for a future world model. (I must quickly acknowledge that others are thinking along somewhat similar lines.)

The building blocks for the global model should be longer-term national models. The national models could (should) have widely varying structures to reflect the different economic structures, levels of development, size, data base, etc. These models should be used to parameterize highly simplified regional models that would be used to merge the nations together into a 'reasonable' number of regions. Finally, a vastly enriched trade structure should link the various regions. I hope that the simplified regional structures would avoid both the potential combinational problems, as well as some of the ones that apparently confronted Project LINK.

Of course, my simplified suggestion skips over the question of how to select the regional grouping. One approach might be to apply principal-component analysis (factor analysis) to the set of nations against a set of dimensions consisting of the key parameters in the simplified models plus a set of quantifiers for items not normally quantified, such as geopolitical relations, religious and cultural values, treaties and agreements, etc.

One may find that there is a class of different regional schemes, depending on which key parameters one selects, which themselves may depend on the purposes of or clients for the proposed model. It would be interesting to compare the results obtained from different, yet logically consistent regional patterns. This approach could bring both a more formal basis for regionalization, and one that might in itself yield new insights into global dynamics.

Comments (by O. Rademaker) on question 4(c). How should changes in structural relations be considered?

The word *structure* has different meanings in different disciplines. As may be deduced from the responses to the questionnaire, it makes sense to distinguish, as attributes of a mathematical model, its variables, the (non)-existence of any direct relations among these variables, and the mathematical formulation of (inter)relations (for example by means of functional expressions, coefficients, etc.) – or, to formulate the list more compactly, the variables, the structure, and the functions (or relation formulae). This list comes down, of course, to definitions of these terms.

An illustration: Roberts presumes that most modellers are not likely to suppose that a Saturn/Uranus conjuncture 'caused' a particular event in world history. If his presumption were credible, it would imply that – even though the positions of Saturn and Uranus with respect to their colleague Terra would be among the variables in the model – the *structure* of the model would indicate the absence of any direct influence of these planetary positions on the model variables related to the 'particular event' in question.

In principle, the distinction between the existence and the non-existence of an influence *in a model* is unambiguously clear, as is any structural change in the form of the (dis)appearance of an influence. Scolnik probably refers to such abrupt structural changes when he writes 'One can conceive that for certain extreme values of certain variables the structural relations will change through a competitive process among the elements of a set of alternative structures.' Meadows also considers such structural changes: 'In some cases model relationships are added or subtracted by means of exogenous policy switches. An entire feedback-regulated pollution-control sector, for example, can be activated or not at the desire of the model operator.' It should be added that such changes can also be made at predetermined points of simulated time and that many programming languages provide very flexible means of implementing more sophisticated *conditional changes in model structure.*

In the real world, the existence or non-existence of an influence is not nearly as clear and as sharply defined as in a model. Further, the modellers decide which influences they include in their model and what they omit, which comes down to 'the influence may exist in reality but we do not expect it to be sufficiently relevant'. A related situation arises when an influence does exist in the model, but does not have to be considered because its effect on the whole is not sufficiently relevant.

Not only the structure counts, but also the relationship functions and their contribution to the behaviour of the whole system or model under study. Most of the responses to question 4(c) refer to this kind of situation.

For example, Meadows writes: 'The entire model structure evolves continuously through time because of the prevalence of nonlinear relationships. The relationships between food and mortality can shift, for instance, from being strongly dominant in the system when food is very scarce, to being one of several equivalent mortality determinants, to being ineffective when food is in abundance'. Changes in structural relationships of the nature considered here and related to the MIT world models were investigated, among others by Hugger and Maier, and by Cuypers, Rademaker, and Thissen.

Roberts clearly refers to the same kind of changes when he writes: 'Changes in structural relationships can be modelled by the inclusion within equations of terms which are not activated until particular features emerge in the external environment', or, as formulated in his answer to question 4(a), 'However, it should be noted that some mechanisms lie dormant and are aroused only by the appearance of specific stress'.

Likewise, in all the other non-linear world models, changes may take place that may be interpreted as changes in structural relationships.

Note that the distinction between the *a priori* and the *a posteriori* situation is relevant. *A priori,* the structure is the determining attribute; the importance of the various functional relationships embedded in the model structure may be evaluated only a posteriori.

Comments (by O. Rademaker) on question 4(d). Should a model be set up as a one-time venture or should it be designed in a modular fashion, flexible enough to adapt to new insights and/or new policy questions? How can such flexibility be ensured?

A simple summary of the responses will suffice.

Linnemann: 'The first model is bound to be a "pilot" model. The modular structure is of course an attractive one but it can only be set up once the basic structural design is constant. After a few years of group experience, such a design may however exist.'

The FUGI team, in their response to question 6, stress the point that 'global modelling requires enormous continuous efforts' and emphasize 'the long-term character of the problems and the need of continuous efforts for revising models'.

Mesarovic: 'The model constructed has to be flexible. That would require the model to have two features: (a) it has to be modular so that any particular aspects of the global situation which has to be dealt with in more details in reference to a given problem area can be represented within the model sufficiently disaggregated; (b) the model ought to be interactive so that a full spectrum of scenarios can be analyzed efficiently. In this connection, we prefer to talk about a policy-assessment (or scenario analysis) system rather than a model.'

Roberts: 'Experience suggests that it is most unwise to assume that a model can be successfully constructed as a one time venture without any modifications arising from new insights and/or new policy questions. Therefore the advice should always be to design in a modular and flexible fashion.'

Scolnik: 'A global model should be modular and flexible and should also change with time. Flexibility can be ensured through a very careful structural design.'

However, Meadows considers the modular, general-purpoe model a mistake:

> So much time and effort is involved in making a credible model that one is tremendously tempted to keep on using it once it is completed, adding to it and adapting it as new information or new problems come up. Some kinds of models for some purposes can indeed be used in this way.

However:

> For large-scale, long-term social problems, I believe the modular, general-purpose sort of model... quickly becomes a monster, enormously complex and incomprehensible, even to its makers. It loses two of the main advantages of formal modeling, transparency... and openness to criticism. Because of the state of understanding of social systems, its relationships are bound to be speculative and subject to much discussion. Believing the results of such a model is an act of faith that no wise policymaker is willing to commit.

Thus there is a conflict between the role of a model as an explicit hypothesis to be comprehended, tested, and perhaps discarded vs. the understandable desire of modelers to see their labors result in widely useful products with long lifetimes. I would suggest that one way to reduce the conflict is to regard the output of modeling to be general ideas and concepts rather than formal devices that deliver precise prescriptions for action. With the more general goal of improvement of the mental models of decision makers rather than the substitution of formal models for decision makers, much smaller simpler models could be made. Each model could be focused on a clear problem definition rather than dealing with all problems. The labor per model would decrease, and then the willingness of modelers to see their hypotheses disproved and to come up with new and better hypotheses would increase. Modeling would become less of a mystical art and more of a science.

Session 5: Testing Models

Rapporteur and Commentator: S. Bremer, International Institute for Comparative Social Research, Wissenschaftszentrum Berlin, Federal Republic of Germany

In response to the question of why more model testing has not been done, one participant suggested that it was not always in the interests of the modelling group to test its model, especially when the group is closely tied to a client, and that the members of the group may not be capable of detecting and testing the deep structural assumptions in their model. Another participant pointed out that one often develops such doubts about the value of a model during its construction phase that one loses interest in extensive testing and goes on instead to build a new improved model. A third opinion was that the insufficiency of model testing may stem from a lack of experience in the global modelling community.

The allocation of responsibility for model testing also prompted several contributions. One modeller stated that having outsiders test one's model was both a painful and rewarding experience and suggested that a model should be offered for testing to those who stand most to lose from its conclusions. Another participant suggested that a modelling group should include an independent individual who has responsibility for testing the model as it develops. The model-testing experience of econometricians was discussed, and it was noted that tests done by outsiders can be very helpful, but that outsiders may inadvertently do violence to a model.

It was argued that one of the reasons that test results are not more widely shared is that the results often yield an understanding or feel for the model that is intuitive and not easily expressed in words. It was suggested that the pressure be put upon modellers to share the insights they drive from model testing.

The participants were somewhat divided about the adequacy of current methods or model testing. Some stated that more powerful methods than those being commonly used were available now, while others were less sanguine about the available methods and called for new developments.

One participant reported on the experience of economic modellers with respect to testing models comparatively. The organizational arrangements whereby this testing is carried out were described, as well as past and ongoing work. It was pointed out that one of the conclusions of this work has been that similar models may yield very different results.

Several participants addressed the question of whether sensitivity is or is not a desirable property of a model; it seemed generally agreed that such a judgement is problem specific and depends on the system being modelled.

One participant disputed the hypothesis that events further away in time are harder to predict than those nearer; long-term trends may be easier to predict than short-term fluctuations.

The suggestion that stochastic factors be introduced into global models was questioned, since to do so we need to make assumptions about the distributions of errors in our data, an area about which we know little.

The larger question of the purpose of testing models was also raised. It was suggested that the purpose of testing must be related to the purpose of the model. The objective may not be prediction, but rather a better understanding of the basic dynamics of the system, for example, and testing then serves the purpose of providing the modeller with a 'feel' for a model. Others argue that our models must be prepared to 'stand up in court' and that testing is the only way we can be reasonably certain that they will.

Several participants stated that the testing phase should be made an integral part of the modelling process, rather than being treated as an afterthought.

The problem of insufficient financial resources for model testing was a major theme that permeated the discussion. Several modellers reported difficulties in securing support for the testing phase of their modelling efforts, and it seemed generally agreed that financial resources are a critical problem.

Comments (by S. Bremer)

These comments are intended to serve three purposes:

1. to synthesize the positions and summarize the activities of the various modelling groups vis-à-vis the part of the questionnaire concerned with model testing,
2. to evaluate these positions and activities with respect to some higher criteria,
3. to provoke some thought about model testing and the broader aspects of the global modelling enterprise.

I recognize at the outset that these are not wholly compatible objectives. The first requires formulating descriptive, objective, and comparative statements about what the groups have said or done. the second entails rendering scholarly and balanced judgements about these same activities, while the third purpose calls for issuing controversial declarations and posing provocative questions.

Given their partial incompatibility and a desire for brevity, I shall not be equally attentive to each of these objectives. The reader of this volume has in his hands statements of the modelling groups, thus obviating the need to restate them. Evaluations are necessary but, especially in this case, imperfect because so much that the groups have done, and their reasons for doing so, are invisible to those outside the groups. The ultimate value of these comments may therefore reside in their ability to provoke thought and discussion.

Before I proceed to the larger task at hand, however, I should make a few preliminary remarks about some definitional and procedural points.

With respect to global models, I will not use the concept 'test' in the narrow, statistical sense of assessing the truth or error content of a postulated theoretical structure, but rather in the broader sense of experiment or trial. Hence, each run of a simulation model is a test within the meaning embraced here.

I will refer to three different types of model tests – sensitivity tests, postdictive tests, and predictive tests:

- When we experiment with a global simulation model in order simply to discover how it behaves under various starting conditions, parameter values, or structural specifications, we are conducting tests of its sensitivity.
- Postdictive tests are runs of a model intended to reveal its ability to replicate past behaviour; they necessarily entail comparing the model outputs with observed data.
- Runs of a simulation model intended to forecast future events and evaluate the impacts of alternative policies on these events are predictive tests.

It should also be noted that I have taken the liberty of rearranging the order in which the topics appear in the questionnaire. I will begin with the narrower, more technical subject of sensitivity tests and proceed upwards through the questionnaire to the larger questions of confidence and uncertainty.

- *Sensitivity tests*

I have defined a test of a model's sensitivity as a run of the model where some starting condition, parameter value, and/or structural specification is changed in order to determine how the model's behaviour differs from some baseline behavioural profile. I sense that this definition is considerably broader than the one modellers employ in general; therefore, I should spell out some of its implications.

A model may show sensitivity to changes in its starting conditions, that is the values of its state variables at time zero. Measurements such as total population and the number of females of child-bearing age are examples drawn from the demographic sector. At time zero we generally assume these to be given and almost invariant, forgetting the sometimes substantial error margins that surround such measurements. Thus, it may be advisable to determine a model's robustness with respect to alternative initial values.

The second form of sensitivity that models may exhibit stems from changes in the parameters, and by common usage this is the most prevalent meaning attached to the notion of model sensitivity. Since we know that, even with the best of data, there is a limit to the precision with which we can estimate the parameters, we would like to know whether changes within this limit alter the model's behaviour substantially.

By structural specification, the third form of sensitivity, I mean the postulated theoretical linkages of which the model is composed. Many modellers may not accept the notion that changes in these linkages fall within the meaning of sensitivity, arguing that changes of this type constitute specifying a new model and not an alternate formulation of the same model. I include such changes, however, because their purpose is the same. If several competing theoretical explanations exist of how some phenomena are interconnected, it is good to know whether the overall behaviour of the model is strongly affected by our choice of one rather than another.

If we employ this broad definition of sensitivity and examine what the global modelling groups have done, we find some sensitivity testing with respect to parameter values, a little with respect to starting conditions (often to smooth out initial irregularities in the model's behavoiur), and almost none with respect to structural specification. Considering the groups individually and judging by the materials at hand, I see them as follows. The World3 group appears to have conducted the most extensive set of sensitivity runs, and they have reported a large number of results. In addition, this model has been analyzed by a number of other scholars. Next in line there are SARUM and Bariloche, the former having used Monte Carlo techniques on some parameters as well as step-change and frequency-response tests, while the latter group has done some classical sensitivity testing of submodels and, more interestingly, applied a more sophisticated optimization technique. The MOIRA group appears to have run its model under differing assumptions about certain key factors that are exogenous to the model. To judge from the available materials, the remaining two groups have not performed much sensitivity testing, or have not yet made the results of these tests widely available.

In my judgement, it would be fair to say that, on the whole, the modelling groups have not devoted much time to sensitivity analysis in comparison to that devoted to the other phases of modelling work. I hasten to add that I do not see this as a serious criticism, since there are a number of good reasons why this should be so.

First, it should be remembered that all of these modelling enterprises are quite young when viewed from the perspective of the history and philosophy of science. Hence, it may be premature to judge them on this dimension.

Second, we should note that the features distinguishing global models from other models make them particularly difficult to test for sensitivity. Their size, complexity, and often nonlinear nature render the classical full-factoral-design strategy inapplicable and make the fall-back position of assuming away higher-order interaction effects untenable. The single factor that may be pointed to as simplifying the sensitivity-analysis procedure is that the models do not yet include stochastic factors (although perhaps they should).

Third, I believe that we lack the proper techniques for assessing the sensitivity of complex models and that our ability to design and construct them has outstripped our ability to analyse them. In spite of progress in the field of research design — and without claiming any technical expertise in this area — I sense that no substantial progress will be made until some rather major advances in artificial intelligence are made that will permit us to automate the testing procedure with the help of powerful heuristics.

Fourth, it is not clear what it means for a model to be 'sensitive' or 'insensitive' to a particular factor. Although we often hear that a model is 'sensitive' if 'small' changes in inputs produce 'large' changes in behaviour and 'insensitive' when the opposite is true, the meanings of both 'small' and 'large' are not self-evidently clear. It may be that 'small' changes are those lying within our personal confidence intervals or zones of indifference; what one modeller views as 'small' may be seen as 'large' by another.

Fifth, I see the question of whether sensitivity is a desirable or undesirable property of a model as fundamental but unresolved. To put the question differently, how much sensitivity is desirable in a model? Models that are relatively insensitive have the advantage that their prognostications can be shown not to depend critically on questionable assumptions, but such models do not serve readily as vehicles for evaluating non-radical policy alternatives. Relatively sensitive models are more able to serve this latter function, but their very sensitivity may, coupled with our uncertainty about how the world works, lead us to assign little credibility to the outcomes of such experiments. In the end, one's attitude towards sensitivity testing and what can be gained from it is probably a function of how one views the world itself. If one thinks of the world as a relatively closed system where inertia and habit operate in powerful, overdetermined ways, then one's model should mirror this relative insensitivity. If one sees the world as an open system with ample opportunities for innovation and creative human intervention, then one's model should have the requisite amount of sensitivity.

When we consider all these factors and add to them the pressure we all exert and feel to make our models even more complex, which in turn makes their sensitivity testing more complex, perhaps we should acknowledge that we lack the resources to do the job in the way we all know it should be done and seek consolation in the assumption that, at this stage of global modelling research, it is appropriate for sensitivity tests not to have high priority.

- *Postdictive tests*

The ability of a model to reproduce past behaviour is widely accepted as one way to enhance the credibility of a model. It is also widely conceded that this ability is a necessary but not sufficient condition for validity. Just how much credibility a model gains by demonstrating postdictive power is, however subject to much dispute. One argument against postdictive power as an element of validity that I will not discuss at length rests on the assumption that the future will necessarily be different from the past, leading to the conclusion that postdictive power says nothing about predictive power. Stated in the extreme form, the assumption is that the future is *independent* of the past. If this assumption is true, then not only must we concede that postdictive power has nothing to do with predictive power, we must also conclude that all efforts to predict the future are equally futile.

Although critics of global modelling have sometimes expressed this extreme view, I think few members of the global modelling community would do so. Most, I think, would conclude that a model's credibility is enhanced by tests demonstrating its ability to reproduce past behaviour. The central question concerns the amount of credibility gained. I suggest that we should assign a model more 'credibility points' based on postdictive tests,

1. the more independent the model is from the data against which it is tested,
2. the longer the time series the model can replicate,
3. the more numerous the independent or quasi-independent aspects of reality the model can reproduce, and
4. the better the fit between the observed values and the behaviour of the model.

These criteria are not to be seen as additive, but rather multiplicative, since a 'zero' on any of the four dimensions indicates that no real test of the model has been made.

The first guideline stems from the fundamental concept of falsifiability. If, for example, we test a model against the data base that was used to estimate its parameters, or if we calibrate or 'tune' some aspects of a model based upon data that are later used to assess its postdictive power, we have 'contaminated' the model, which therefore cannot be falsified by these data. Many books dealing with simulation techniques exhort us to 'save' data in order to preserve some for testing purposes, but everyone knows that it is extraordinarily difficult to do so when data are scarce. Hence, the dilemma that all global modellers face: use all of the data, probably produce a better model, or save some of the already insufficient data, probaby produce a poorer model, but retain some basis for evaluating it.

The second criterion stems from my hunch that the longer the span of time a model is able to replicate, the more successfully it reflects higher-order dynamic processes. By higher-order dynamic processes I mean such things as evolutionary, self-regulating, self-reproducing, adaptive processes that tend to operate and be observable only over long spans of time. The incorporation of such processes is very important if we wish to make long-term predictions. A rule of thumb concerning spectral analysis that was once suggested to me by a statistician may be relevant here. He suggested that a time series should be *at least* three times as long as the longest cycle that one wishes to be able to detect. In the present context, this suggests that a model that can replicate the period 1900 to 1975 should not be asked to predict beyond the year 2000. Of course, rules of thumb cannot be proved valid; we can only judge whether they are useful.

The third of my four criteria deals with the number of independent or quasi-independent aspects of reality that the model can reproduce, and some clarification of what I mean is in order. I start from the position that the more phenomena a theory can account for, other things being equal, the better the theory; since simulation models are theories, in my view, the same thing applies in the case of global models. Hence if two models can successfully replicate the time path of the gross national product, for example, but one can also replicate its component parts (consumption, investment, government, and trade balance), I would assign the latter model more credibility. Incidentally, this is an example of what I mean by quasi-independent aspects of reality, since the components of the gross national product are linked by an identity relationship.

The fourth criterion, goodness of fit, seems to be the least controversial of all, but in some ways it is not. There are those who feel that, for a variety of reasons, only qualitative assessments of goodness of fit are necessary and, perhaps, possible. A qualitative assessment involves comparing the shapes of time paths of variables traced by the model and observed in the referent world, and no statistical procedures are applied to the data. With this approach it becomes very difficult to discriminate between models, given the essentially subjective nature of the 'eyeballing' process. For example, to the eye, both World2 and World3 postdict world population during the period 1900-1970 equally well. Is it not reasonable to ask which one reproduces the growth of population during this period with greater accuracy? If this is not a reasonable question, then how do we justify making a submodel more complex if a very simple version will capture the shape of the time path? On the other side, those who want to see the whole battery of statistical goodness-of-fit techniques applied may forget the limited accuracy of most of our data and the restrictive assumptions of many statistical tests.

Turning to the question of what the various global modelling groups have said or done with respect to the subject of postdictive power, we find the following. The creators of World3 begin the runs of their model in 1900 and devote a portion of a chapter in their technical report to analysing the behaviour of the model between the starting date and 1975. They look at several aspects of the model's behaviour, but employ only qualitative judgements about goodness of fit; a careful observer will note some small anomalies in the model's behaviour — for example a rather precipitous drop in the death rate in the middle of the period. The main problem is, of course, that we are not given any real-world data values against which to compare the model's behaviour and are, for the most part, merely told that the model behaves in accordance with the overall trends found in the referent world. They assert that it is not difficult to make a model retrace a time path if one is willing to give up the constraint that relations and parameters make some theoretical sense.

The MOIRA group also seems to hold the position that matching historical time series is not difficult to achieve if one is willing to sacrifice all the degrees of freedom, violating my first criterion. A review of the available MOIRA documentation reveals, however, that they consider the simulation of historical development paths as one step in their methodology, and, although they report no such runs, they do make brief reference to the comparison of model-generated tariff policy results with real-world data covering the period 1966-1971.

The SARU group acknowledges that postdictive power is a necessary but not sufficient criterion for validity and feels that testing a model against historical data can reveal weak structures in the model. In a recent study they attempted to replicate the impact of the 1973 oil-price rise; they appear to have had some success, but replicating longer periods of time would require changing the starting conditions for their model, which are derived from data representing the late sixties and early seventies.

The Bariloche group claims that its model can replicate the time paths of several variables between 1960 and 1970, but it is not clear how many time points are actually present in the test data. In some cases, one suspects that only the 1960 and 1970 values are observed, those in between being missing or interpolated. In addition, they employ a built-in calibration procedure, which appears to adjust the coefficients of the five production functions in every run. Thus, it is difficult to say how much credibility these tests contribute to the model.

With respect to the FUGI and M/P models, I am unable to say what kinds of tests of postdictive power were carried out or are planned, but it would appear that the M/P group, believing that many models can be made to reproduce the past through the 'tuning' process, may also feel that tests of postdictive power are not very exacting ones.

What surprised me most about the responses of the groups is the near consensus that replicating the past is not a very exacting test for a global model. I strongly question this position and doubt whether any existing global model could demonstrate much postdictive power if the four criteria outlined above were adhered to. Consider the first criterion, which stipulates that the model must be independent from the data against which it is tested. A good test would require, for example, that we estimate the model's parameters on one set of data (perhaps 1900 to 1940) and then, without using additional information, ask it to predict values for another period (1941 to 1975, for example). I frankly do not think any existing model would fare well in this exercise, nor would any that seem likely to be developed soon. And if the rule of thumb suggested earlier that three years of good postdiction yields one year of credible prediction is reasonable, then our job seems formidable indeed.

This would mean that, if we wanted to predict events between 1975 and 2100 with some confidence, we should be able to postdict the period 1600-1975. Selecting the year 2050 as our time horizon for prediction would require postdicting the period 1750-1975, and if we are merely interested in developments to the year 2000, our model should be able to postdict the period 1900-1975.

I conclude by saying that, although it is not easy for a model to predict the past when reasonably high scientific standards are used, in the short run, to demonstrate such ability may be the only way to discriminate between better and poorer models.

- *Predictive tests*

The modelling groups were asked to reply to the question: 'How can one be ensured that the model will correspond to reality?' I take this to mean: 'How can we be sure that a model's predictions about the future are accurate?' The obvious answer is we cannot be sure, and the modelling groups seem to share this view.

Let us rather start with the question: 'What is it that global modellers are trying to predict?' Everyone seems to agree that global modellers are not trying to make pinpoint predictions about anything; in some cases, it appears as though they seek to make ball-park predictions about everything. The original mandate of global modelling was to explore the 'predicament of mankind' and seek solutions to the 'world problematique', neither of which really forces us to be very specific about what it is we should be trying to predict. In the meantime, a certain consensus seems to have emerged around a problem nexus that includes population growth, resource depletion, food supply, standard of living, and environmental quality. Some modellers, such as the World3 group, attacked all of these problems simultaneously, while later efforts, such as MOIRA, chose to focus on fewer, or even only one, of these problem areas.

In spite of this consensus, it is very difficult to compare the predictions of the various models. First of all, there is not a great deal of overlap about what they predict, although population growth and food supply appear to be two promising areas for comparisons. Even where such an overlap exists, we often find that their selection of different measurement units for the same variable, for example food, makes such comparisons very problematical. In addition, they partition the globe in different ways, which means that, at best, their predictions are comparable only at the global level. And, finally, their time horizons vary significantly, ranging from the beginning to the end of the twenty-first century. This leaves us with perhaps three to four decades of potentially comparable predicted values.

I believe also that it can be said that the modellers exhibit different predictive styles. Oversimplified and overstated they appear to be:

1. World3: what will happen unless something is done soon.
2. Bariloche: what could be realized if something were done.

3. SARU: what is likely to happen if the global system continues to work as it has.
4. MOIRA: what policies are likely to lead to a better-fed world.
5. M/P: what social adjustments and political decisions need to be made in order to achieve global equilibrium.
6. FUGI: what economic developments are consistent with other economic developments.

Although these are, in some sense, caricatures of the models and their purposes, they indicate how the models differ in flavour with regard to their predictions.

In a more fundamental way, however, the predictions of the various models *are* similar. Almost without exception they assume *implicitly* that certain fundamental political problems will somehow be solved, and they assume *explicitly* that certain important political problems will somehow not arise. With respect to the first of these assumptions, I see, for example, the World3 and Bariloche groups both implicitly assuming a radical alteration in the distributive mechanisms that determine who gets what. If one examines, for example, the equilibrium solution offered by the World3 group and superimposes upon it the present inequalities between and within nations, one discovers that approximately 25 per cent of the world's population will receive less than 60 dollars' worth of industrial output per capita. And, although the Bariloche group recognizes that maldistribution is the fundamental problem, their model assumes, rather than telling us how to achieve, an equal distribution of the basic needs within regions. And regionalization itself assumes that the world is reduced from 150 or more jealous and competitive decision-making units to a dozen or less, a transformation that would substantially reduce the political complexity and anarchy of our present world.

Among the important political problems that are likely to arise, but which are not dealt with by the global models, one has to list international and civil violence. And, while recognizing that it is necessary to concentrate on some aspects of reality and neglect others, I consider it a serious shortcoming that none of the models under consideration addresses the problem of the sudden and drastic reduction in life expectancy that a large portion of the globe's population may experience owing to international and/or civil wars. If we recall that in the last 150 years there have been about a hundred international wars and approximately an equal number of civil wars, which resulted in deaths ranging from a few thousand to many millions, it seems that a simple extrapolation of past experience would lead us to expect a substantial loss of life from this source in the next 150 years.

Why is it that global modellers have, almost without exception, excluded such phenomena as wars and revolutions from their foci? Some may feel that these are relatively unimportant in comparison to such things as famines. Granting their importance, some may feel that wars and revolutions are not likely to have much impact on long-term global trends. Perhaps some see such phenomena as one of the many things that may result from the forecasted global shortages, rather than as causal agents in the creation of such shortages. I do not find any of these arguments persuasive. More often one hears that such phenomena, although important, are 'hard' or even 'impossible' to predict. In response, I ask whether such phenomena are really harder to predict than others, and whether this is the main criterion that we should employ in selecting our predictive focus.

Let's consider for a moment some tentative generalizations on the subject of what should be 'hard' and what should be 'easy' to predict. Other things being equal:

1. Events near in time are easier to predict than those farther away.
2. Events heavily dependent on prior events are easier to predict than those relatively independent of prior events.
3. Events in some parts of the world are easier to predict than those in other parts of the world.
4. Events occurring frequently are easier to predict than those occurring rarely.
5. Events flowing more or less inexorably from major events are easier to predict than those triggered by relatively minor developments.
6. Relatively familiar events are easier to predict than unique ones.

One could list many more abstract characteristics which, in an intuitive way, seem to be shared by phenomena that we feel are hard to predict, but the important point is that, when viewed in this way, many 'natural' disasters, such as droughts and epidemics, are no easier to predict, in principle, than 'unnatural' disasters like war and revolutions.

Against this assessment of difficulty of prediction, we need to balance what I call the 'need for prediction'. Tentatively, I equate the need for predicting a phenomenon as a combination of these components:

1. the probability of its occurrence,
2. the speed of its onset,
3. the magnitude of the disturbance it causes,
4. the irreversibility of its effects,
5. the persistence of its effects.

If we take all of these considerations into account, I believe that we can conclude that, as difficult as global violence may be to predict, it is imperative that we do so.

• *Confidence and uncertainty*

'A chain is only as strong as its weakest link', the old adage tells us; to a certain degree this summarizes one point of view about global models. This view is incorporated in a tactic that is rather frequently used by critics of global modelling: focus on one or two of the many relations that a model usually contains, argue that they are absurd, unsupported, or speculative, and conclude, therefore, that the entire model is absurd, unsupported, or speculative, since a model is only as good as its weakest component.

The chain metaphor does have some merit when we examine a particular model. It reminds us that it is extremely unlikely that all of the components are equally strong and suggests that the demands that we place on the model must be tempered by an awareness of the location and strength of the weakest components. The metaphor also suggests two strategies for dealing with weak links: we can replace or reinforce them, and thereby, with a relatively small amount of effort, increase the carrying capacity of the model substantially, or we can divide the chain at its weakest link and produce two shorter but stronger chains. In the modelling enterprise, the latter strategy is equivalent to uncoupling certain sectors of the model and using them for limited studies.

The chain metaphor has its limits, however, and there are those who argue that a model is not at all like a chain, but rather a network of interacting and reinforcing factors. Some of the components may be very weak, but they do not affect the overall strength of the model. In other words, the model is stronger than its weakest component, and one cannot measure its strength by examining its components.

Of course, some models are more chain-like than others and some portions of models are more chain-like than others. Even in networks there often are critical components whose failure can render entire systems useless. Global modellers need to examine their models and assign levels of confidence to different parts of their models by a combination of objective and subjective means.

As I read the responses to the questionnaire and what documentation I have at hand, the positions of the various groups seem to be as follows: World3 and MOIRA seem to place greatest confidence in the general theory or structure that the model is based upon, although this similarity is severely reduced by the different levels of generality of the two models. Bariloche and SARU seem to be more confident in the parts of their models that are empirically grounded. The M/P group feels most confident in the parts of their models that incorporate processes that change slowly and incrementally. It is difficult to see where the FUGI group stands on this question, although I suspect that, stressing as they do the quality of the data, they base confidence assessments primarily on empirical considerations.

When we turn to the topic of what parts of their models the groups have the least confidence in, the weak links, as it were, we find World3 acknowledging some difficulties in the economic sector, while Bariloche admits uncertainty about the quality of the international trade process and the values of certain parameters that specify maximum values or rates of technological change. MOIRA indicates that they are most uncertain about factors that are exogenous to the model. In addition, their representation of the centrally planned economies is admitted to be of poorer quality than that of the market economies. SARU states that they are most uncertain about the relationships that had to be estimated subjectively; an examination of their report reveals that these are essentially concerned with time functions, which control such things as smoothing, rates of erosion and mobility, and the like. Little can be said about the weaknesses of the FUGI model, although, since it incorporates an input-output table, the commonly noted problem of assigning future values to the input-output coefficients must be a source of uncertainty. The M/P group expresses reservations about the assumptions related to value choices.

On the whole, I believe the modelling groups are to be congratulated for their frankness in admitting weaknesses and accorded praise for the parts of their models in which they express confidence. Furthermore, I take it almost as a law of modelling that, except in rare instances, no one knows the strengths and weaknesses (particularly the weaknesses!) of a model as well as those who developed it. It seems imperative to me that this knowledge be communicated to us, so that we can evaluate the model and, if we choose to improve it, locate quickly the parts of the model most in need of revision.

I see three ways this can be accomplished. First, we can communicate informally, largely at conferences of the type that IIASA has organized. Second, we can include in our technical documentation objective and quasi-rigorous judgements about the strengths and weaknesses of various parts of our models. Third, we can use and develop ways of making uncertainty an integral part of the structure of the model. By the systematic introduction of parameter-governed stochastic factors, for example, models could be run in a deterministic mode, as they are now, or subjected to varying degrees of uncertainty. Although I see many problems associated with this third strategy, I believe we need to take some steps in this direction.

• *Conclusion*

In conclusion, I would like to pose five questions that stem from my remarks.

1. *Are we satisfied with the amount of model testing that has been done?*

I think most of us conclude that the answer is no. Among the modelling groups that have been surveyed here, I would say that the World3 and SARU groups have done the best job of model testing – or at least the best job of communicating test results. I place Bariloche and MOIRA in the next group, and the M/P and FUGI modelling efforts in a third. It appears in general that, relative to the other activities required in a global modelling enterprise, global modellers have not devoted much time to testing their models. This may be an erroneous impression, since model testing is more or less a continuous process, many of the results of such tests are not publicly available, and such results may not be easily retrievable.

2. *Why hasn't more model testing been done?*

A variety of answers may be given to this question. It could be that, after one has invested years in developing a model, one loses the enthusiasm and interest that are necessary to see the model through the testing phase. Or, if the modeller's enthusiasm has not waned, perhaps that of his sponsor has, making it no longer possible to maintain the research group as an ongoing enterprise. These are practical explanations for insufficient testing, but the truer reason may lie elsewhere.

3. *Who has the responsibility for model testing?*

Most global modellers would probably respond initially that it is primarily the responsibility of the modelling group itself. I suggest that the community of researchers to whom the model is addressed share this responsibility, and, if the scientific enterprise is viewed from a certain perspective, I think it can be argued that the primary responsibility lies, not with the modelling group, but rather with the research community. Many philosophers of science dispute the conception of science that calls for the theorist to subject his own theory to falsification efforts rigidly and ruthlessly, arguing that this has not been an essential ingredient in scientific progress and is not logically necessary for such progress. What is necessary, however, is a research community composed of scientists engaged in similar endeavours that subjects the ideas of its members to close and careful scrutiny. Thus, one answer to the question of why more model testing has not been done may be that the relevant research community does not exist, is too small, or lacks the desire to do the job.

4. *Why haven't more comparative tests been carried out?*

One specific form of test that has been conspicuously absent so far is the one that treats the models, or their components, as competing theories. In spite of the problems of comparability that I outlined above, I see some opportunities in this area. One could, for example, extract relatively comparable sectors from different models and run them under similar sets of conditions, observing how much their outcomes differ and why. This approach might be feasible in the population and food sectors. A second, admittedly more difficult, alternative involves interchanging parts of different models. If we approach the problem creatively, I am sure other revealing tests can be devised.

5. *How can we make test results more widely available?*

I assume that there is more model testing going on than is commonly known, and I suspect that one of our key problems is disseminating the knowledge gained from these tests. Perhaps we need some type of coordinating institution or mechanism whereby the global modelling community can from time to time assimilate and evaluate its collective effort. IIASA has in the past, through its sponsorship of the global modelling conferences, taken a major step in this direction; the hope that they will continue to do so is widely shared.

Session 6: Organizing Modelling Work

Rapporteur and Commentator: J. Richardson, Quantitative Teaching and Research Laboratory, College of Public and International Affairs, The American University, Washington, D.C., USA (the Mesarovic/ Pestel model)

The sparse attendance at this session indicated that social-system modellers are not overwhelmingly interested in examining the professional social systems of which they themselves are a part. Those who participated, however, did have a high interest in the subject of organizing interdisciplinary modelling efforts. Although the hour was late, the discussion was unusually lively and candid. Much of it consisted of shared personal experiences of working in modelling groups, global or otherwise.

Two conclusions were almost immediately apparent. First, global modelling groups face exactly the same problems as any interdisciplinary research groups, with perhaps an added emphasis on the difficulties of problem definition (it is seductively easy to try to do too much with a global model) and relations with the public (great public interest, especially in the first few models, raises difficult questions of when and how to communicate results). The second conclusion is that the conference participants were reluctant to draw conclusions. Different groups seem to have followed quite distinct styles of interaction, leadership, task and resource allocation, and documentation, most of these styles seemed to be accidental rather than deliberate, and no one was willing to make normative statements about how a group *should* be organized. Insofar as any general lessons are to be drawn, it is apparently up to me as originator and evaluator of this session to draw them. The comments that follow are my attempt to do so.

Introduction

Where team efforts are involved, research reports are, almost inevitably, pale reflections of the richly textured social processes by which research is produced. While engaged in the process of developing global models, team members have had little time, and probably less inclination, to examine the process of research introspectively. After the fact, there is even less incentive to do so. (The only exception of which I am aware is found in Jørgen Rander's doctoral dissertation. (A.P. Sloan School of Management, MIT, Cambridge, Mass., September 1973). Randers quotes extensively from a diary kept during the evolution of a model he was developing for the National Council of Churches. Most modellers who are familiar with this (unpublished) work characterize it as a good description of their own modelling experiences). The literature dealing with research management has little to offer in the way of specific guidelines for organizing such projects. It is doubtful, moreover, that those actually involved in the projects had much familiarity with such guidelines as do exist.

There is a need for introspection, however. I believe there must and will be a growing number of projects that in significant respects resemble those discussed at the Sixth IIASA Conference on Global Modelling. In particular, such projects are likely to:

- deal with complex systems that cannot be fully understood from the perspective of a single discipline, or paradigm;
- require a time horizon of twenty years or more;
- require a global perspective;
- deal with issues falling at the interfaces of politics, economics, the natural and biological sciences, and technology;
- deal with issues of interest and concern to policy makers about which decisions are being made during the course of the project;
- be characterized by a commitment to a systems approach, in some form, and to modelling.

The projects reviewed in this conference are of particular interest because they must, in some degree, be regarded as 'successes'. To be specific, each of the projects:

- secured leadership;
- recruited and organized a multidisciplinary team and managed, more or less, to keep it together and functioning throughout the project;
- defined a more or less manageable problem (or set of problems);
- collected data and organized it sufficiently to contribute to the process of model development;
- wrote equations describing various sectors of a model;
- integrated these sectors into an overall structure;
- wrote a computer program that reflected the equations;
- got it to run;
- produced output that could be presented as 'results';
- incorporated these 'results', as well as other materials, in various reports and other publications for presentation at IIASA and elsewhere.*

---

*Of course, by the time a project was selected as the focus for an IIASA conference, there was a pretty good indication that *some* results would be forthcoming. But there was not, in every instance, the certainty. In some cases, an impending IIASA conference became a motivating force, leading to near superhuman efforts to produce and organize results in a presentable form.

Moreover, several projects have proved to be quite successful in retaining or securing long-term support (MOIRA, SARUM, FUGI). Even the projects whose long-term survival seems less certain have been active for several years beyond the first IIASA presentations and attained a high degree of visibility.

The queries and the responses

The queries to the modelling groups on the topics of internal organization were directed principally to three issues:

- the internal structure, management, and leadership styles of the project groups,
- the special problem of coordination between subgroups having responsibility for different submodels or substantive areas,
- the 'external relations' of project groups with fellow global modellers and with the wider scientific community.

The perennial problem of documentation was also included under the 'internal organization' category.

Although the queries principally solicited recommendations (example: 'How should global modelling work be organized?'), it was hoped that responses would draw upon actual experiences with regard to internal organization and management.

As expected, responses differed in depth, substance, and emphasis. The two groups that described their own experiences talked about an organizational structure that evolved, based on experience. Several groups started as 'shoestring' operations with funding becoming available only after some initial results had been produced. Surprisingly, the roles of the project leaders were scarcely mentioned, making it difficult to support (or refute) my own belief that leadership and management styles are very important factors in successful projects.

Nothing startling was reported on the issue of coordinating subgroups. In general, project groups got together on a weekly or monthly basis. More communication was required where several subgroups were engaged in submodel development (Forrester/Meadows, Mesarovic/Pestel) than where the actual modelling was the responsibility of one or two modelling specialists (Moira, Bariloche). Apparently, subgroup coordination did not pose (or is not remembered as posing) serious problems for any of the projects.

The question of cooperation with other global modelling groups was viewed as even less consequential. Most groups either felt that cooperation had been adequate or felt that more interchange, while desirable in principle, would have diverted time from other more important activities. IIASA Global Modelling Conferences and meetings organized by the Club of Rome were most frequently mentioned as places where groups could get together to exchange ideas and information. Although it was not mentioned in the responses, I think it is worth noting that relations between global modellers have become decidedly more cordial, cooperative, and mutually supportive in recent years. Perhaps the best evidence of this is the type of open discussion possible at the Sixth IIASA Conference. During the other years, at least some project members viewed other groups as competitors for scarce resources of funding and political attention (and treated their members accordingly).

No respondents felt that global modelling was sufficiently unique to require extraordinary approaches to interrelating with the wider scientific community. Rather, there was a clear consensus that presentations at scientific meetings and especially publication of results in scientific journals were the appropriate media for communication. At a more informal level, modellers were urged to communicate with colleagues from the more traditional scientific disciplines.

Documentation

This subject merits a special heading because, increasingly, it is viewed as a serious and urgent problem for the field. Donella Meadows describes the problem with characteristic candor:

> Global models, above all, should be good examples of the modelling art and should be accessible for discussion in a wide variety of arenas. That means that their documentation should be (1) clear to both analysts and laymen, (2) easily available, and (3) timely. Not one global model, including our own, has met all three of these documentation conditions; most have met none of them. This lack of documentation is a disgrace. No other supposedly scientific discipline would permit such irresponsible reporting.

Part of the documentation problem reflects the personal interests and styles of most modellers. Seeking the apogee of perfection that awaits 'just one more modification' and 'just one more run' is an exciting, albeit frustrating, enterprise. Preparation of documentation is tedious, not at all exciting and incredibly time-consuming. Moreover, many modellers who might like to document their work properly are simply not trained to do so.

But rigorous, clear documentation requires educated clients who insist upon it as well as modellers who are capable and committed. Presently, many clients and potential users seem uninterested in documentation and overeager for quick 'results'. Developing policy-oriented models poses special problems in this regard, since the timing of a particular recommendation to meet a felt need is of such critical importance. In the long run, it is inevitable that a mature field of global modelling, if it develops, will be characterized by adherence to the standards of documentation advocated by most project groups. Individual modellers can help the process along by personal adherence to high standards, combined with candor and unflinching integrity in describing their own work. In my judgement, the documentation of the World3 model (albeit belated) and of the SARU model provide excellent examples.

Internal organization and 'success'

It is of interest to observe that many of the factors assumed to contribute to the success of interdisciplinary research activities have not been generally characteristic of the global modelling projects. For example:

- Only three of the projects had a specific client; at least two, during part of their lives, were 'bootstrap' operations, which functioned with limited and uncertain funding.
- Several of the projects were characterized by lack of consensus regarding the operative paradigm, and even by conflict among team members regarding paradigmatic issues.
- Few of the projects seem to have been guided by a carefully defined and well-developed plan (although most had a clear conception of overall objectives). Many decisions were made on an ad hoc basis.
- No particular discipline or paradigm seems to have been particularly influential, overall, although economics seems to be growing in influence. This is partly due to the fact that many policy makers are unaware of alternatives to economic modelling. It is not due, in my judgement, to any demonstrable superiority of economic models for long-term global problems.
- The projects have differed significantly in the degree of involvement of project members in actual modelling. In the Meadows/Forrester and Mesarovic/Pestel projects, all of the core group members were directly involved in the technical aspects of modelling; there was relatively little in the way of a methodological division of labour – except that in the Mesarovic/Pestel project, the distinctive MODEL BUILDER software was developed separately. At least two of the other projects (Bariloche and MOIRA) numbered among their teams one or more 'modelling specialists' who were responsible for model-development activities and served an integrative function.

Factors contributing to 'success'

One searches in vain for a substantial number of characteristics, common to all of the projects, that might have contributed to their 'success'. I have been able to uncover only two:

1. High motivation is clearly the most important common characteristic and one that distinguishes the global modelling projects from many other efforts. Virtually all of the 'core' team members have had a deeply felt belief in what they were doing. This made possible real sacrifices and a level of productivity that would have been inconceivable under other circumstances.
2. Each of the projects with which I am closely familiar has had a 'core' group of three to five persons who functioned with a relatively high level

of involvement and integration, if not harmony. These members were available, more or less on a twenty-four hour basis, to meet commitments, take up slack, and deal with emergencies.

Problems and issues

In this section I list some of the major problems and issues that have faced global modelling projects. Although few of them were mentioned by the teams, I believe they are significant and merit further discussion.

1. *Leadership strategies*

The obvious generalization about the leadership strategies that governed the project is that no generalization is possible. Project leaders differed in their backgrounds, personal objectives, and styles of management. It would be valuable for each leader to describe the basic management principles that guided him during the course of the research. Perhaps there are areas of commonality not apparent on the surface.

2. *Recruitment and personnel management*

Recruitment to global modelling projects and to 'core' membership seems mostly to have been based on self-selection and ad hoc considerations. But there were many potential candidates for project team membership, including candidates with excellent qualifications, who did not become closely involved. It would be interesting for each project group to describe the process by which the team coalesced, as well as its degree of stability and patterns of interaction.

3. *'Timing'*

If any projects had a PERT chart or other device that played a significant role in the timing of key activities and decisions, I am not aware of it. However, a clear sense of timing, especially for various 'external-relations' activities was one of the most important contributions of several project leaders. For example, making predictions about when a set of reasonably presentable 'results' would be forthcoming (so that commitments for a presentation could be made) was a task requiring skill, courage, and luck. Often, the life of a project depended on the correctness of such decisions.

4. *Delegation of responsibility and division of labour*

As noted above, projects have been characterized by different patterns of organization, both with regard to substantive and methodological tasking. Also, the degree of involvement of project leaders with the technical details of programmes, submodels, and results has differed significantly. We have no real understanding of the factors that contributed to these differences or their consequences for the projects.

5. *When and how to 'go public'*

I would guess that every project has agonized over this decision. 'Going public' changes the complexion of a project's activities, especially when the audience includes the general public. Often it has involved making commitments that could not easily be retracted and, subsequently, led to a 'defensive' posture vis-à-vis critics and other global modelling projects. On the other hand, a willingness to 'go public' — to make visible statements about important issues — is the essence of policy-relevant research. In more recent global modelling projects, there has been a reluctance to tackle the really crucial issues facing mankind. I believe that this shift in emphasis has been motivated, at least in part, by an awareness of the problems involved in 'going public'.

6. *Popularization: when and how?*

A closely related issue is whether or not a project should attempt to produce a work designed for the general public, similar to *The Limits to Growth* and *Mankind at the Turning Point*. Apparently, the later projects have not considered popularization worth while. I think this is unfortunate. Certainly, many others who have written about some of the long-term development issues facing mankind have been less competent to do so than global modellers. I would be interested to know whether or not projects considered such a work and rejected the idea.

7. *Interaction with policy makers*

All of the projects claim to have produced policy-relevant research. However, the approach to interacting with policymakers and the resources devoted to this activity have been quite different. Identifying and explaining 'successes' in this area would constitute a particularly valuable contribution.

Conclusions: personal reflections

My selection to evaluate the responses dealing with internal organization and to write this summary was not accidental. For almost fifteen years I have grappled with the problem of organizing multidisciplinary research activities of one sort or another. At global modelling conferences, I have regularly pleaded for a degree of self-consciousness as the several projects developed and evolved. In informal discussions with project leaders, I have probed for operative principles of management that might provide guidance to others.

At the Sixth IIASA Conference, it seems reasonable to ask, 'What does all this concern with project management and internal organization add up to?' The answer is that very little of general relevance has been uncovered. Each of the projects has been, essentially, a unique event. Somehow, in particular circumstances, small groups of individuals came together who combined, in sufficient degree, leadership ability, vision, modelling talent, public relations capabilities, an incredible capacity for work, and, most important, entrepreneurial talent. The several models that emerged, the 'results' published, and the documentation (or lack thereof) were produced by the synergy of such groups.

It would probably be foolish to try to prescribe for this sort of project, because the essential characteristics of leadership, commitment, talent, and so forth, may appear among individuals possessing and requiring very different managerial styles. But it is also unnecessary to prescribe, because the types of projects motivated by the Club of Rome and first presented at IIASA are only characteristic of the nascent stages of an emerging issue area or field.

At the Sixth Global Modelling Conference, there was considerable discussion about the future of the field — were we witnessing a 'rising or setting sun'. The evidence suggested that the concern with global issues is a growing one, that modelling will continue to play a significant role in exploring such issues, and, more specifically, that several of IIASA conference projects appear to have good long-term prospects. But if the Sixth Conference marks the continued emergence of a field, I also believe it marks the end of an era. I doubt there will be additional major projects characterized by the idealism, the (sometimes unfounded) confidence, or the excitement (and occasional despair) that are part of pioneering and the creative process.

We must now get on with the more mundane activities through which art becomes craft and (where appropriate) craft becomes applied science.

Session 7: Modeller-Client Relations

Rapporteur: J. Robinson, Management and Technology Area, IIASA, Laxenburg, Austria
Commentators: J. Robinson, IIASA; B. Hickman, Department of Economics, Stanford University, USA

The discussion of modeller-client relations in global modelling efforts showed that the subject looks very different from different perspectives. The fragmented, multireality character of the discussion can be displayed better using dramatic than essay form. Accordingly, I break with convention and summarize the discussion with a little plotless play. I have tried to preserve the content of discussion, but have fictionalized the speakers by aggregating people who seemed to be saying the same thing, by splitting people who seemed to be saying two different things at once, and by making speeches more terse and blunt. The intent is to avoid associating points of view with individual speakers, thereby running the risk of emphasizing and deepening the conflict which has at times divided the global modelling community. Instead I have associated points of view with life positions. The reader, accordingly, is encouraged to look upon the speakers as masked figures, as in Greek drama, speaking from typical life roles rather than as individual known personalities.

The text, like the discussion, is neither conclusive nor orderly. It proved beyond my skill to bring the material into a coherent, self-consistent format without taking sides. I was tempted to manipulate the text to make the idealistic, anti-mercenary position win, thereby counterbalancing the opinion set forth in my evaluator's remarks, and I have tried in the text to strengthen the idealistic position. However, I coud not go further without converting the supposed summary into a polemic. Therefore, I offer the material as it came and leave it to the reader to search out logical order or definitive conclusions.

Moderator

I have three points to make. First, I hope this discussion will generate light, not heat and muddy abstractions; therefore, I ask speakers to address global modeller-client relations on the basis of their own experience, rather than on the basis of their beliefs and ideologies. All of us can reel off our beliefs, many of us disagree with or disapprove of one another's beliefs. If we steer discussion in the direction of belief, I predict we shall have a lovely, empty argument. Therefore, please, let's restrict the discussion to facts, to experiences. The literature is poor in observing what modellers and clients have done and in observing what practices have resulted from what approaches. I suspect we can all learn from one another's experiences, particularly if they are examined honestly and self-critically.

Second, I hope we can hear from clients as well as modellers. What little discussion there has been of modeller-client relations has been almost entirely from the modeller's perspective. Clearly, every relation has at least two sides, and it seems unlikely that relations can be improved unless all sides are taken into account.

Third, I foresee a definitional snarl centred around the word 'client'. In the usual sense of the word, client may refer to the provider of funds, or to the user, or both, but does not usually refer to interested onlookers, critics, and the general audience. I would like our discussion to include all three classes to whom the modeller must relate: funding agents, users, and audience. All three classes are important to the future of global modelling: funding agents for its economic viability, users for its direct effect on policy, and audience for both peer review and criticism and indirect effect on policy through influence on popular values, norms, and perceptions of the future. I suggest that we use these more specific terms in place of the word client to avoid semantic confusion.

Global modeller

I want to act like a real scientist, but the world is not tidy, like physics. My conscience wouldn't stay out of the laboratory, not when the fate of the globe and humanity is uncertain. So I took up social-system modelling. Clients, funding agents, and audience are the bane of my existence. I hate to be bought, but my work requires $100,000 a year – more really – to do good work. I need intelligent criticism, but most of my critics simply don't understand my model. As for users, I'm not sure they exist – apart from politicians who use the model when it substantiates their positions and disregard it when it shows them wrong.

Corporate user

Multinational Business Inc. is in the market for a clear, practical model that will help us understand how we fit into the international picture. It is worth millions to us to play the system right. The model must give a credible representation of MNB affairs, keep a global perspective, and be simple enough that the Directorate can understand it.

Idealist

'He who pays the piper calls the tune'. But we're not talking about a dance or a nightclub. We have serious responsibilities. The global models we build affect the global future. From our models come society's expectations. From our expectations come our plans. From our plans comes the future itself. We must act responsibly. For example, if we stress in our models the ideal of meeting basic human needs for all the world's people, we help swing the power balance to the side of the poor. There are clients for such work – the United Nations, some Third World nations.

Manager

The Ivory Tower is a very nice place, but unless someone takes care of the practical details, things will come to a grinding halt. We need global models that are oriented towards practical ends. We need to know if inflation will get worse, whether LDCs will default on their loans, and what will happen to the prices of basic commodities. Of course we care that people are starving. But starvation will only get worse if practical business is handled without foresight. Modellers must come down to earth and begin to work on real questions. Good intentions don't do anything unless they are hooked into a working system. Unless models have real-world clients, all they will produce is utopian solutions, recommendations that no one on earth can carry out, so divorced from reality that no sane person would try to carry them out even if he had the power to try. Academics need the discipline of serving practical men. I don't care whether it's in business or in government, model users must have a hand in model creation if the product is to be useful.

Second manager

Success and failure will often depend on how the modelling effort is organized. One sure recipe for failure that we must avoid is third-party models. When B commissions C to build a model for A it almost invariably happens that A has no use for the model. Many analysts seem to think the path to success lies in working with high-level politicians or with top management. This is a delusion. Top management seldom has the time and often lacks the training to direct a modelling effort personally or to sort through model findings trying to ascertain their real-world implications. A better strategy is to use an intermediary, a management analyst or policy analyst, to direct, sort through, and evaluate analytic findings. Realistically, global modellers are going to have to learn to work through such intermediaries if they want to have an ongoing, established influence on government or corporate policy.

Global modeller

What, honestly, do you do when your work has not proceeded to the point where you can believe your results, but they are asking for answers? Asking? Hell, they are paying. Do I write the report for them, glossing over the uncertainties, or even worse, fooling myself into believing my work, despite the flaws I know are in it? If I resist doing this, how do I finance the basic work I know needs to be done?

Scholar

Quality, I'm sure, will win out in the end. Shoddy work will not stand the test of time. No, I am not so naive as to think we can make an objective science of it. But there have been false claims and poor documentation; they have given the business a bad name. We don't help anyone if our work is lousy.

Global modeller

My experience with the public sector has been bad. They pay you to tell them what they want to hear. They distort what you tell them so it supports the positions they have already taken. The price goes up if the results are politically useful. It is better to sell your soul to a corporation than pander to politicians.

Third World analyst

All models contain ideologies. Most of the global models are anti-natalist. That is what sells now in the West; clients buy the models that correspond to their ideologies. Malthusian models permit you, the rich, to absolve yourselves of blame, to say our poverty is our own fault because we have too many children, to justify not adjusting prices and giving aid, to forget that Western imperialism supports corrupt and backward regimes. The game can be played in other ways. It is as easy to create a structure that shows that equitable income distribution is essential to economic growth as to build one that shows birth rates are too high. We must build — and publicize — models of equity to keep the debate on global futures balanced. Yes, funding is scarce. But audience and users are plentiful. Most of the world's population is made of poor Third World people. Stop worrying about funding. Where there is a will there is a way.

Observer

Why do we need more global models anyway? Aren't six or ten, or however many there are, enough? I'd rather see a few of the existing models cleaned up, documented, tested, and improved than to see more and more models built. How many times are we going to invent the wheel? Is it going to get so that each analyst has his own personal global model? I vote for some sort of agency to go through the existing global models and decide which would be useful to build on; thereafter the modelling community should channel its energies to improving these, rather than to litter the field with more half-baked projects.

Comments (by B. Hickman)

Most replies did not answer the questions on this topic. The following tabulation of replies to the individual questions includes either direct answers or, in some cases, remarks not explicitly responsive to particular questions but having a bearing on them.

1. *Should an explicit client be identified at an early age?*

Asdela – No, since global modelling should neither be a commercial activity nor be oriented to limited interests of individuals or particular groups.

Mesarovic – In models for policy analysis, the input of the user is more valuable the sooner it is received in the process.

Roberts – There is no general answer to the question of whether an explicit client should be identified. A client may be necessary in order to secure funds or to obtain approval for the work to be done. The modellers may believe that their work is only of value to particular decision makers, in which case it is clearly essential to work to specific clients. Conversely, it is clearly possible for the results of modelling to be of value to the world at large, in which case there is no special client and the customer can be regarded as mankind itself.

2. *To what extent do clients participate in the work?*

Asdela – Our project was financed by the International Development Research Centre of Canada, which did not participate in the work at all.

Meadows – For policy-related projects, let the client define the problem; listen carefully to him, involve him regularly in the model's development. However, it is doubtful if these general guidelines apply to global modelling, since clients who are really responsible for long-term issues of global scope are almost non-existent.

Mesarovic – The relationships between the modeller and user should be as intimate as possible, regardless of whether a global or local model is concerned.

MOIRA – The project was a Club-of-Rome initiative. No specific instructions were given and funding was by the Dutch Government to maintain scientific independence.

Roberts – In the case of SARUM, approval had been provided for the development of a model, but without specific questions, the answers to which were required.

3. *What formats to communicate with the clients should be developed?*

Meadows – Always explain the model with clear language, graphs, charts; avoid mathematics and professional jargon.

Roberts – All possible means: text, visual aids, general presentations, private dialogue, interaction exercises. Metalanguages may be necessary to assist understanding of dynamic relationships by users with limited knowledge of mathematics.

4. *What can help the client to avoid misunderstandings, to understand the strengths, weaknesses, and limitations of the model?*

Asdela – It is essential to explain clearly and without mystifications the hypotheses, virtues, shortcomings, etc., of the model. If the user is requested to provide exogenous variables for defining scenarios, he should be aware of the implications of his decisions, of empirical knowledge abut these variables, and of the reasons why the variables have not been made endogenous to the model.

Meadows – Involve the client regularly in the model's development; if he does not understand the model and see his own assumptions about the system in it, he will never implement it.

Roberts – Ultimately the client has to take a part of any modelling exercise on trust. If a client were able to pass beyond this stage and not to require any trust at all, this would imply that he (or she) would have become a member of the team itself. The trust that clients should place in modellers' work ought to be derived from the modellers' track record. In modelling areas other than that of the globe, there is some evidence that this advice does apply. The most robust arrangement for advising clients derives from an open field in which modellers with different structures and starting from different viewpoints can offer their goods freely. Criticism is available from members of the peer group as for scientific work. Under these conditions there is far less risk of clients misplacing trust.

5. *Which clients need world modelling most? Which are most interested in world modelling?*

Asdela – The UN agencies, some governments, some academic groups.

FUGI – It is worth noting that our global modelling efforts have been financially supported, not only by our own budget but also by other organizations such as NIRA (National Institute for Research Advancement) and UN ESCAP (the United Nations Economic and Social Commission for Asia and the Pacific). Some other policy makers also would be interested in our global model, so that it might help to identify internationally concerted actions for growth of national economies, both at the regional and the global levels.

Meadows – Clients who are responsible for long-term issues of global scope are almost non-existent. In a sense, the clients of global modelling are all of humanity, or at least the section of the world population that is sufficiently educated and free from poverty to think about the long-term future.

MOIRA – Only funding agencies should be considered that are already aware of the limitations of global models. Then the modeller is not tempted to make golden promises or oversell his results. UN agencies, governments, and universities should be the main clients.

Roberts – The findings of world modellers are likely to matter most with respect to changes in attitude. Thus, people (as such) are likely to be the most important clients from modelling. The group that is most interested in world modelling appears to be the senior staff of large companies.

6. *Do you have recommendations on how to reduce 'overselling'?*

Asdela – If the previous recommendations about relations with the user are followed, then 'overselling' will be reduced automatically. However, this is extremely difficult to achieve, because those who oversell models are the first in not following these kinds of recommendations. On the other hand, models should not be sold as 'magic' black boxes, instruments of a new form of intellectual colonialism. In this sense, it is necessary to encourage the formation of autonomous groups in several countries, having the capacity of criticizing and comparing different models and constructing their own ones. Perhaps we could think of creating an ethical committee of global modelling to help solve these problems.

Roberts – The phenomenon of 'overselling' is self-correcting. Overselling is followed by disillusion and a lack of receptivity to work that follows (even though the later work may be superior to what was oversold). This is unfortunate for the purveyors of work that is of high quality and is not being oversold.

● *Comments on the replies*

A common theme is that long-run global models are most appropriately sponsored by international agencies, governments, and universities. This is because the models deal with problems concerning mankind as a whole, and hence it is essential to maintain scientific objectivity and independence.

With regard to involving clients or users in model construction, some replies suggest that such involvement might vitiate the scientific independence of the project. Is this necessarily the case? What are the costs of not involving the users during the modelling stage? Are global models so remote from policy actions that it is unnecessary to involve the policy maker in model formulation to gain his confidence in the model?

Modellers are concerned that users understand the limitations of the models as well as their strengths. Full disclosure of critical assumptions and weaknesses is obviously a necessary condition to achieve this end. Is there enough agreement among modellers about the appropriate validation procedures, however, to meet the prior condition that the modellers themselves understand the limitations of their models? Can competition among modellers provide the necessary safeguards in the absence of agreement about validation criteria?

Comments (by J. Robinson)

Global modelling started as client-directed, under a strong client, and shifted to become academic and client-independent. The client-centredness of the first global models is clear. The Club of Rome set the ball rolling – first they hired Hasan Ozbekhan, a cybernetician, with the hope that he could assist in conceptualizing and clarifying the tangle of interrelated problems that the Club called the 'global problematique'. The Club remained in a directing position; when Ozbekhan's work stayed too long on too high a level of abstraction, they withdrew support; whereupon Jay Forrester appeared on the scene.

Forrester is by training an engineer, and he related to the Club in a fashion typical of engineer-client relations. After ascertaining the problem focus of the intended model at a Club of Rome meeting, Forrester built a simple working prototype model, World1, for client review. World1 was enthusiastically received, and was revised and documented as World2. After approving the prototype, the Club found funding for a full-scale marketable product, World3, which was constructed by former students of Forrester.

The Club only relinquished control when the exercise had produced their desired product, as evidenced by the fact that client pressure drove the modellers to violate their scientific standards by publishing *The Limits to Growth* before the technical documentation for World3 was completed.

- *The modeller as client*

Later global models have become progressively less client-oriented. The Mesarovic/Pestel world model (otherwise known as the world integrated model or WIM) originated under the special circumstance of having the modellers among their own clientele. This allowed the modellers to follow their proclivities with minimal external control. The result was a large detailed model with central conceptualization designed in accordance with the modellers' perception of the global system. In theory the model was user-oriented. However, for the Mesarovic/Pestel model, 'client-oriented' had a distinctly different meaning than for the three World models. In the World models, 'client' was a specific, named entity. In the Mesarovic/Pestel work, 'client' was a generic concept, and the model aspired to be adaptable to many clients – mostly policy makers. The World models were custom designed. The WIM was not. In my opinion the multiple-client strategy has been only partially successful. That is, a large number of agencies have been convinced to contract for use of the WIM and have reflected with interest and curiosity on its outputs, but it has not to date come to be well understood or institutionalized as a planning tool by any of its users.

Like the Mesarovic/Pestel model, later global models were based on modeller perceptions of the world and have sought clients for their work in national governments and international organizations. The UN world model,

FUGI, SARUM, and the Latin American world model have all been brought into policy analysis in one way or another. None of the modellers involved are terribly enthusiastic about the results of their work in the policy arena, and several are openly cynical. Complaints against policy makers include the following: they only listen when you tell them what they want to hear; their time horizons are too short to use long-term models; and they are not open to insights about how the world operates.

- *Will the client believe?*

The policy makers' version of these stories remains unrecorded; however, given the large number of people with solutions to the world's problems and the equally large number pointing out the difficulties with the policy makers' solutions, it is not surprising that policy makers have not paid global modellers much heed — especially given the controversial character of many models and their typically large number of unrealistic caveats (for example assuming no wars or revolutions and no technological shifts). There may be a basic principle to be discovered here. For example, where the part of model structure clearly relating to client interest is dominated by a structure that does not relate to client conceptualizations, it is very difficult for the client to understand or trust the model. If this rule holds, it follows that models have to be almost custom designed to establish client rapport, and that modeller-conceived multiclient models are prone to failure.

- *Get rid of clients?*

The logical next step after multiclient modelling is 'clientless' modelling, meaning not modelling without a sponsor (which, given the costs involved, is almost impossible) but modelling where neither the sponsor nor parties associated with him are involved in model conceptualization or development.

The only obvious niche for clientless modelling is pure science, where funding is procured for theoretical research and research findings are to be presented to scientific peers. Pure science, however, is an uncomfortable niche for the global modeller. First, competition is tight and funding has become extremely hard to secure. The scientific establishment appears to question whether global modelling is a science — or an exercise in ideology — and is not easily convinced of the value of further global modelling work.

Second, the traditional modes of communication in the sciences are not adequate for the presentation of global models. Scientific journal articles are a poor forum for the description of models. Model documentation is difficult and expensive to prepare, voluminous, and boring. Publishers are hard to find, audiences are sparse. Once published it is as likely as not that an article will merely gather dust. The only fast way for academic articles to reach a wider public is for them to be picked up and transmitted by the popular press —

which may sufficiently distort an article's content that the publicity becomes more of a curse than a blessing.

Third, the academic atmosphere, with its selection pressures favouring the esoteric, the refined, the theoretically pure, and the specialized, is not a particularly healthy environment for multidisciplinary global modelling. Criticism from an academic audience does not generally help a modeller develop an overview of the real world, or assist him in figuring out what in his model is of general importance, what is unclear or incomprehensible, and what is nonsense. Rather, it tends to call his attention to abstract details, such as whether he would have done better to use some other data base, computing routine, or estimation procedure.

Ironically, the attempt to be purely scientific and client free seems to be responsible for giving global modelling something other than the basically cooperative and universal-minded spirit that is supposed to characterize pure science. With funding scarce and audiences growing deaf, global modellers are apt to make a case for their own work and deprecate the work of others, only to be followed by others who denounce their work in turn.

• *The past and the future*

Over the course of the past decade, global modelling has moved out from under the wing of the Club of Rome and come to be sponsored both as a policy tool and as a form of academic analysis. In general, recent sponsors of global modelling have not known what they want from global models, and have not found much use for their output. Shortages of funds have constrained some modellers and eliminated others. At times competition for the limited available funds has become uncivil and characterized by overselling and slander. Public audiences seem bored with global modelling, and the scientific community remains sceptical.

If modeller-client relationships continue in the present vein, a scenario for the next decade might be:

1. Mesarovic, the Carter-Petri (the UN world model) group, and other multiclient strategy modellers will find occasional commissions by various bureaucracies, but no meaningful sponsorship.
2. The institutionalized modelling groups (SARUM and FUGI) will be funded until some bureaucratic adjustment cuts them off.
3. Two or three more books will come out documenting global models, but they will probably not sell well.
4. Global modelling will pass away as a fad.

The above scenario, like most trend extrapolations, misses turning points; it is unlikely that global modelling will arrive at a dead end. Three factors are operating to alter the present trend:

1. Modellers are becoming more professional.
2. Potential global model clients are getting more numerous and more sophisticated.
3. The global situation is becoming more critical, making the need for models greater and more apparent.

The three factors are examined separately below.

- *Professionalism replacing pioneering*

The early years of global modelling were characterized by exuberance, unrealistic expectations, and inexperienced management. Costs and time requirements were grossly underestimated with the consequence that model construction greatly overshadowed model testing, documentation, and refinement. Much of the energy used in global modelling has gone to 'starting costs' such as learning how to run interdisciplinary investigations, finding data sources, constructing software, and finding capable staff members. Modellers appear to be catching on to procedures such as budgeting for documentation and testing; and start-up costs decrease with experience. Those who continue with global modelling should be able to turn out better quality work at a lower cost than did their predecessors – although the excitement of the earlier work has died down.

- *Changing clientele*

Various institutions – including the World Bank and oil companies – have discovered that world models constructed to solve their specific problems can be useful, and have begun to employ modellers. The resulting models are tending to be custom-designed, as were the World models. There also appears to be a tendency for national planning models to expand their treatment of the rest of the world – and to move towards greater consideration of intersectoral activities. In the United States, for example, the International Energy Evaluation System (IEES) has replaced the Project Independence Energy System (PIES); almost everywhere agricultural models are being expanded to account for increasing energy costs.

Furthermore, clients should be getting more sophisticated. Each year more people are trained in computer techniques of all sorts, and each year people with training in systems analysis gain seniority.

- *Changing global situation*

Finally, the seriousness of global problems will increase over the next few decades, and with it the urgency for new integrated approaches to their management. Modelling could serve a useful role in reducing the uncertainties, and helping to derive efficient and humane strategies for management.

In sum, global modelling is an evolving institution in a rapidly changing environment. In the past, modellers have looked increasingly to academic sponsors for support, which appears to be an evolutionary dead end – both because academia lacks funds and because it does not provide selective pressures that encourage pragmatic, interdisciplinary, holistic thinking. However, pressures are mounting, from modellers and clients, and from changes in the global situation as well, to bring global models out of the ivory tower and into a useful role in dealing with global problems.

Session 8: General Evaluation

Rapporteur: G. Bruckmann, Global Modeling, IIASA, Laxenburg, Austria
Commentator: P. Neurath, Sociology Departments, Queens College, City University of New York, and the University of Vienna

The topics discussed in this session were grouped into five categories:

1. The relation between global modelling and the established science.
2. The present situation of global modelling. What has it achieved? Is it in a period of 'sunrise' or 'sunset'?
3. Future paths to be pursued in global modelling.
4. Concrete actions to be taken to improve the field.
5. IIASA's role in this context.

1. *The relation between global modelling and the established sciences*

During the discussion, several speakers pointed out that it would be fallacious to see the relation between global modelling and the established sciences as an antagonism; on the contrary, cooperation should be strengthened in the sense that the achievements made within the established sciences should increasingly be incorporated into global models. Linnemann went as far as to call this relation a 'non-issue': there is no distinction between global modelling and other fields; there is, however, the danger that auxiliary sciences (like mathematics and computer sciences) will dominate global modelling in the sense that they decide on the content of models. Carter thought it would be helpful to keep apart the fields of global modelling and futures research, which tend to overlap, owing to clients' pressure. In her opinion, the role of global modelling should be to increase the understanding of historical relations before using models to predict the long-term future.

2. *The present situation of global modelling*

Linnemann said it may be true that global modelling is approaching sunset, but one should not forget that after sunset the sun usually rises again. The first sunrise of global modelling would never have been possible without the pioneering work of Forrester and Meadows. The sunset which seems to have occurred after this first period of overexpectation must be welcomed, as it allows global modellers to be more careful in their research and analyses and to be less eager to give quick-and-ready answers to policy makers. This period of reflection is necessary before the next sunrise.

Rademaker added that the 'sunset' of global modelling is shown by the fact that output has been slowing down considerably. This, however, is due to an increasing professionalization of the field; and, as the potential need for global models remains great, the sun will rise again. The future success of global modelling, however, will be a function of its credibility. Commenting on the role of this Conference, Rademaker said it had been useful in providing scientists who usually are too involved in their work a chance to reflect on their research and become more conscious of the framework in which they have been operating.

Richardson agreed that the Conference had been very useful at this present stage of global modelling: it had been characterized by an absence of defensiveness, by a general openness about critical issues, by an increasing degree of understanding and commitment to evaluation and testing, by an increasing degree of awareness about the need to improve standards of documentation, and by an intense dialogue between economists and scientists of other disciplines judging modelling work from different paradigms. Roberts underscored the importance of such a process of self-evaluation; such a venture should be repeated every five or six years. Hickman also welcomed what he called the therapeutic effect of the Conference, providing an opportunity to share the experiences of modellers with a wider audience from other scientific disciplines. A (minor) defect of the Conference, however, can be seen in the fact that not all participants were equally acquainted with the models discussed. He suggested that, in the Conference proceedings, a section should be allocated to a brief description of the models, plus a complete bibliography associated with them and with comparative studies done in this field.

3. *Future paths in global modelling*

As to the future of global modelling, Meadows considered the Conference a positive indicator: the high degree of openness and honesty exhibited by all participants could be the key to solving specific problems of global modelling: problems of micro versus macro, of testing, of environmental variables, of limitations of data, of policy considerations. The spirit of humbleness may give rise to more imaginative and daring experimental approaches, which may be more necessary than technical refinements.

Hickman expressed the view that global modelling might have to redefine its own time horizon. While he agreed with the working definition of global modelling as given in the Conference (comprehensiveness, closeness, and time horizon of more than thirty years) the last item is already undergoing some change. LINK, with its time horizon of two to three years, is not considered a 'global model' by the definition given above. However, there are already several global models in existence that focus on a medium-term time horizon: MOIRA, FUGI, and the Leontief model. It may well be the case that economic models will move up and global models will move down towards a time horizon of ten to twenty-five years.

Moore, after some discussion, gave an explicit list of topics that deserve particular attention by global modelling in the years to come: environmental issues, better understanding of international trade structures, the relation between GNP and energy consumption and the flow of armaments, as well as closing the gap between the policy maker and the modeller by building policy variables into a model. Roberts stressed the importance of the treatment of environmental issues, however difficult this may be.

Richardson added that global modelling will have to deal with certain problems of particular importance in the years to come: a new economic order, the pressures due to population growth, and depletion of resources(energy), resulting in a possible widening of the gap between rich and poor countries. Hopkins expressed the view that it is not so much the problems of tomorrow but the problems of today that are pressing: the world is facing a crisis now with 800 million hungry people. In his opinion, the population growth issue is seen in most global models with an anti-natalist bias, with no differentiation between countries. Also the depletion of oil and coal reserves seems to be a debatable issue. The environmental issue must also be seen in the light of the poor countries, where the poor judge environmental standards differently. For many of these issues, national models may prove much more appropriate than global models, as they get closer to the problem, whereas global models tend to oversimplify. The debate, however, should not boil down to the question of global versus national models, but rather which model is better suited for which issue.

Kile went one step further by stating the need to be better aware of one's own cultural, philosophical, or religious background: market economies, centrally planned economies, less-developed countries will arrive at different approaches to the same problem. There are many possible approaches not represented at this Conference at all.

After several speakers had commented upon the need to include policy considerations, Linnemann proposed to integrate policy makers into the team of global modellers to be able to achieve better results. One method is to include advisory boards of ministers or businessmen in the team; thus, freedom of action is maintained, but 'policy making' explicitly included.

4. *Concrete actions to be taken*

Several speakers expressed the same view as Linnemann that concrete actions should be taken to close the gap between the policy maker and the modeller; to provide policy variables in a model may not suffice.

Besides this problem, Moore suggested three types of measures to be taken to ensure better communication between global modelling groups:

(a) the exchange of models in physical form (for example by tapes),
(b) a journal explicitly devoted to global modelling, and
(c) the need for a common language of communication in order to be able to compare models.

Dickhoven drew the attention of the audience to the fact that his work concerns primarily Moore's points (a) and (c). If models are to be regarded as planning tools, they must be consolidated. The model-bank system for the Federal Republic of Germany is almost completed; it will include some of the most important models for administrative purposes and a set of additional instruments useful for analysing models and data collection. A service centre for modelling work has been established, both for the internal use of its institution (Gesellschaft für Mathematik und Datenverarbeitung) and to administer the use of the models. It will also serve educational purposes in applying the models, and it will allow the scientific community to make comparative studies of different models. If this work should be broadened upon an international scale, more resources would be needed; if other centres elsewhere would pursue similar work, a well-planned division of labour should be made to avoid duplication of effort. He invited anyone interested to use the facilities at GMD as a basis for discussion with researchers, practitioners, and other users of models.

Coming back to the question of an appropriate journal, Walker said that a *Journal of Policy Modeling* had been founded by the Society of Policy Modeling in New York. Scolnik added that not only this journal but also two others could serve global modelling: the *Journal of Applied Mathematical Modelling* and *Human Systems Management*. It would be more desirable, however, to have a specialized journal.

5. *IIASA's future role in global modelling*

Several speakers stressed the importance of the IIASA global modelling conference series and urged its continuation. Federwisch suggested a conference on international economic models; several participants expressed the view that the treatment of environmental problems within global models should receive the most urgent attention. Linnemann thanked IIASA for the impetus the IIASA conference had given to the MOIRA project; a similar impetus can be expected from any future IIASA global modelling conference.

Scolnik stressed that IIASA, in addition to holding conferences, could play a considerable role in documentation, and also by publishing a newsletter on global modelling. Meadows seconded Scolnik concerning the importance of finding an institutional solution for the documentation problem: it will be of great importance to have some clearing house for documentation and for fixing standards to compare models. Such a function could also be served by several institutions if one institution, for example IIASA, assumes a leadership role. The success of an institutionalized solution, however, will largely also depend upon the individual commitment of every global modeller.

Richardson added that IIASA has a unique opportunity in its management and technology group to focus on these issues from a methodological point of view. Tomlinson replied that the IIASA management and technology group is definitely interested in examining techniques and models as they can be used for policy making. Recent developments have added a lot of sophistication, not only in global modelling but also in gaming, but little has been achieved in the use of such models as policy-making tools. He agreed that knowledge and experience have been collected by some industries, like oil companies or the automotive industry; he said that IIASA is investigating ways of adapting models to policy making.

Asked to summarize IIASA's position, Roger Levien the Director, made these remarks:

> Let me refer to the implicit definition of global models used in this meeting. They have been taken to be models that are global in geographical extent, that are multisectoral, and that have a long time horizon. I was gratified that no one in the audience urged IIASA to develop a model of this type. Indeed, I did not hear anyone indicate an intention to begin a new development of another such global model. I think that is an important unstated conclusion of this conference. Perhaps it demonstrates the awareness among this group that the flurry of initial activity and the blaze of publicity are dying down. The real questions this group has been facing are: What will remain? What should the next step be?
>
> One alternative, which does not seem entirely improbable, is that nothing will remain. Many of the groups who have already constructed global models seem to have turned their attention to smaller, more tractable problems and to have contracted their horizons. The other alternative is that something does remain, that the groups settle down to the hard scientific and professional work of understanding what seems to me to be their real topic: global interrelations and interdependencies. This would reflect the recognition among these groups of the hubris of trying to develop in one step a theory of global interrelations. Now humility has replaced hubris, in recognition of how many steps must be taken to reach the goal that many thought was within their reach. So the question that faces us, and it certainly faces IIASA, is what can be done to make progress along this path? The hopeful answer is that what should be done is to discipline and professionalize the study of global interrelations and interdependencies.
>
> Now I want to ask whether the emphasis should be on modelling, which is after all a tool, or on the subject itself. And I want to ask whether we should rush to application – to policy recommendations – before we gain a deeper level of understanding. I shall express my own prejudices: we ought indeed to continue in a disciplined and professional way to try to build our understanding of global

interdependencies and interrelations, both geographic and sectoral, and we must enhance that understanding greatly before rushing to conclusions. In gaining understanding, modelling is an important technique, but the emphasis should be on the substantive understanding and not on the technique.

In this line, then, there are three topics of interest for IIASA. The first is building substantive understanding of geographic and sectoral interdependencies. In some ways, everything we do addresses this issue. We have thus far addressed it from the sectoral viewpoint, studying first the energy and food sectors. But now we are looking specifically at the linkages, the mechanisms for interdependency among sectors and regions. The economic mechanisms are our initial interest, but certainly there are many other mechanisms of which we ought to try to gain a disciplined understanding.

Second, there is the question of improving the tools of modelling, including such things as improving documentation, validation, verification, standards, testing, and so on. This is important, not only for global models, but for all other policy-relevant models. Much of what IIASA does involves modelling of one sort or another, and for our own sense of professionalism, we have to work on improving these tools. Thus, both for reasons arising from this review of global models, and because of the extensive role of modelling at IIASA, I think we should try to establish an international focal point for improving the craft of modelling.

The third way IIASA interest is represented is by this series of conferences. The Institute seeks to facilitate communication – international collaboration, if you wish – among groups facing common problems. And certainly, since we have all felt that this series of conferences has been a successful one, we would like to find some way of continuing to encourage such exchanges. However, it need not be through conferences of exactly this form. There are many other mechanisms; some of which might be better.

We have now had six global modelling conferences. They have served a useful role at this stage in the evolution of global modelling in the international arena. But before committing IIASA to its next step, I think we must consider very conscientiously the most fruitful way to proceed. For example, our pest management work, which is not conducted at IIASA any more, has catalysed the formation of a pest-management network. This is a group of about 150-200 people who are interested in systems analysis applied to the problems of pest control and who have a formal mechanism of exchanging preprints and holding working meetings through IIASA. It is conceivable that such a network might be established for the group of persons concerned with global modelling, with IIASA serving as a catalyst and facilitator. But I do not want to limit my consideration to these suggestions; there are other mechanisms that might be proposed, and I am open to suggestions.

Our concern is that this mechanism for communication not be continued solely on the basis of its momentum, but rather that the decision be made after we ask ourselves what the appropriate next step is and what function IIASA can best play at this time in its development and that of global modelling.I assure you of the continuing interest of the Institute in this activity and its many ramifications.

Comments (by P. Neurath)

As a sociologist – and not even a mathematical one – I can't help feeling like an interloper at this conference. My reasons for being here are that I am a statistician and demographer, and I have studied and written about your work during the last four years. Still, as an outsider I am observing you and your world in a way that reminds me of a book that described how a three-year-old child experiences the world of the grown-ups. Its tital is *Nunni among the Giants*.

Now, meeting you at last with the naive and amazed eyes of a child, I am making an amazing discovery: the giants in their grown-up sphere, which ordinarily we children dare not enter, play games rather similar to ours. The same joyful playing with new ideas, the same formation of little groups that stake out their own turfs but then cheerfully play with others from other turfs, the same watchfulness about what the outside world is going to say about their play, and to some extent the same drawing together in search of identity as a group when the outside closes in or, worse, when the outside tends to treat them as nothing but little children playing some odd kinds of games that the grown-ups don't quite understand, or care to understand, because it is, after all, only child's play.

I am, of course, overstating the case, because I am fully aware of the enormous impact that global modelling has had (ever since the publication of *The Limits to Growth*) upon the public debate about the issues of population explosion and resource depletion. Even if they never knew and never will know what a global model is or can do, millions of people acquired through the debate about the first model a feeling of possible impending doom — or at least the need to assert that they don't believe in such doom, in spite of what some of the model makers may be saying.

However, I do not intend to talk to you about the social importance or, for that matter, about the theoretical meaning of what you are doing. Instead, I want to make a few remarks in my role as an observer at your conference, that is as 'Nunni among the giants'.

Let me preface what comes next with saying that, in a way, I went through this some thirty years ago at some of the earliest meetings of the American Association of Public Opinion Research and the American Institute of Mathematical Statistics.

A new science, or a new branch within a field of science, or, if you want to call it so, a new specialty within a given field, characterized by a set of methods, techniques, and theories too esoteric for the outsiders to follow fully, has been in the process of establishing itself. And, not too surprisingly for a sociologist, its practitioners, feeling lonesome in their places of work and insecure about their position in the world, are looking for others with similar interests; they come, they meet, they find help, security, and a sense of identity in each other's presence. Eventually, if the history of other cases allows a projection into the not-too-distant future, they will form their own society, with its own journal. I recall at an early meeting of the American Association of Public Opinion Research a formal motion to appoint a committee for the purpose of devising a name for the activity in which we were engaging and, more important, a professional title for its practitioners; the maker of the motion said that it should be a title ending with '-ologist' or '-ician' or 'some such thing'. You already have a name for your subject, I will not be surprised if, within a few years, I will attend a conference of the 'International Society for Global Modelling'. If you don't have the name for your society yet, take it as a free gift from Nunni.

To repeat: as an outsider I am, first of all, struck by the search for an identity among you, quite naturally coupled with a feeling of insecurity about the activity in which you are all engaged, whether it is worth while, whether it will help the world with its problems as you are hoping that it will, or whether what you are producing will be abused by others for power purposes. I also note search for agreement on basic theories, concepts, and methods, so that a group working on a new model will not need to establish its right to make a model before being allowed to proceed with its own special approach to its particular results.

Second, I am struck by the limited degree to which methods and results can be compared from model to model, and by the lack so far of a common language and a common set of concepts and basic canons in terms of which the quality of work and results can be judged. However, you have discussed these points at great length. We sociologists are now in the midst of similar difficulties, so we are not surprised when we see them elsewhere.

Third, I am, to some extent, astonished that so many of you skirt the essentially social, economic, and political nature of your work, of your results, and of the impact that it has on the social and political world around you. True, each of you is fully aware of this. Nevertheless, I find it amazing that so eminently political an activity can be discussed in such generally unpolitical terms.

This brings me finally to a more personal observation: I admire the spirit of cooperation among you and the genuine civility with which you are treating each other, even when there are obvious differences in approaches, results, and implications, including political ones. Too, I admire the dedication and enthusiasm with which you are going about your work. You have not yet become as professionalized as the scientists who started their branches many years earlier. There are still among you many who joined your organizations, not just in search of a job but in search of a place to make a contribution to mankind. Your work is a joy for an old outsider to see.

To which I will add an observation from my position as a demographer: note how few of the people present are of my generation and how many are younger. This is, of course, no accident, but a by-product of both population growth and the most recent industrial revolution. Anything concerned with computers is basically the province of a younger generation with its own technology: hardware, software, and new intellectual activities — of which global modelling is one of the newest.

Welcome to the playground!

6

What have global models taught us about modelling?

If we step back from all the details, the opinions, the contradictions, and the divisions of the conference and view them at arm's length, some basic patterns and conclusions become apparent.

The most obvious conclusion comes as no surprise to global modellers, but may be new to those outside the field who have perceived global modelling as something apart and different. The conversations reported in Chapter 5 could have been carried on by almost any interdisciplinary, international group of social-system modellers. THE METHODOLOGICAL PROBLEMS OF GLOBAL MODELLERS ARE COMMON TO ALL SOCIAL-SYSTEM MODELLERS. Because of the holistic, long-term, urgent, and controversial nature of the problems they work on, global modellers may fall into the traps of modelling practice more frequently than other modellers, but the traps are there for everyone. If global modellers are able to avoid them, the lessons they learn will be generally useful to other modellers of complex social systems.

And lessons do seem to be emerging. The many issues of theory and method taken up at the conference can be organized into two categories. On some issues there was widespread agreement about the best way to proceed, and often the extent of agreement surprised everyone. Here it seems possible to say with some degree of finality, 'This is the way to model'.

On other issues the modellers could only agree to disagree. If any prescription can be made here, it must be to regard these areas as appropriate places for experimentation, for tolerating and encouraging a variety of approaches, and for honest reporting of results. As Michiel Keyzer says of one of these areas, 'Let all the flowers flourish' (an English translation of a Dutch translation of Mao Tse Tung).

This chapter lists the topics brought up at the conference in each of the two categories. In order to end positively, the listing begins with disagreements and ends with agreements. The disagreements are presented by stating the extremes of the argument, to which is then added editorial commentary aimed at assessing the continuity of the positions (if any) between the extremes, the balance point (if any) at which the competing spokesmen can meet, or how to proceed (if there is a way) when the opposing views are firmly entrenched.

Let All the Flowers Flourish: Areas of Disagreement

1. Should models be built to answer a single well-defined question or should they be built to represent many aspects of a system and serve many different purposes?

On the surface, this disagreement is usually phrased in terms of complaints about the way the other side models:

Making general-purpose models results in large numbers of virtually incomprehensible equations that, for all of their complexity and detail, are still unsatisfactory representations of reality. They only give the appearance of answering questions; the answers are bogus.

Making single-purpose models just results in simplistic answers that are often totally wrong because they ignore the crucial complexities of the total system.

Underneath these arguments there is a basic difference about the purpose of modelling:

The purpose of modelling is to make the incomprehensible clear by exhibiting its essence. A system cannot be bounded and its essence cannot be discussed without reference to a particular problem or question. Therefore, clarity cannot be achieved without starting from a clear problem definition.

The purpose of modelling is to use the computer to deal with complexities that the mind cannot understand. The more you can put into a model, the more likely it is to give the proper answer to any one question and to all questions.

Upon deeper probing, both sides usually become uncomfortably inarticulate, which is a sign that slumbering paradigmatic beliefs are being awakened. At issue are not only the purposes of modelling, but also the nature of the world being modelled and our ability to know it. It is worth penetrating one more layer, because the differences here underlie some of the other disagreements we will come to later. (While books could be — and have been — written about these differences, we will only characterize them briefly here.)

The world is made up of deep causal structures that produce the surface behaviour that we observe as strings

The world is made up of sets of distinct systems boundable by similarities in surface characteristics of separate events. By asking clear questions about behaviour, observing carefully, and using systems theory, we can discover the basic structures, which are usually much simpler than the complex event chains. Similar structures produce similar behaviour in widely different systems (for example the same structure underlies the oscillating, goal-seeking behaviour of economic markets, blood-sugar regulation, thermostats, and predator-prey systems). We can never know the full details of the world or answer all questions about it — no computer could hold so much information, even if we could assemble it. But we can learn general behavioural tendencies and probable policy responses.

(national economies, ecosystems, populations, the transportation system). As we learn, we can produce more complete representations of these systems, which will be useful for more purposes. We can even couple these models to represent The System. The models will grow beyond our capacity to understand them, but we can apply continuous tests of their output against the unfolding events to tell us what degree of confidence to have in them. We can never be confident of underlying causal structures, because they are unseen and unseeable and cannot be compared directly against observable events. But we can replicate past series of events, and thereby predict future ones.

It is difficult to hold both sides of this discussion in one's head at one time, and therefore a balance point between them must be very unstable, if it exists at all. We have known modellers to start in one position and, over a lifetime of experimentation, gravitate to the other (both directions) or to oscillate back and forth between them, but never to capture both in one model. Since the two approaches are bound to produce totally different kinds of information (although neither kind is recognized as reliable information by the other side), it is probably to the benefit of the human race to have the argument go on, as long as it remains civil.

2. Should models be made in direct response to pressing issues of public policy or should the goal be general improvement in understanding?

It is simply irresponsible to go sounding off to the policy world on the basis of uncertain, untested models that have not been subjected to the scrutiny of the scientific community. Theory, testing, discussion, and improved understanding should precede policy application.

Decisions are being made daily on the basis of mental models that are even worse than our own — admittedly bad — computer models. It is irresponsible not to present new insights to the world, even if the insights are not complete.

This is the old argument between the scientists and the engineers. In the modelling field it is mostly conducted between the global modellers, who so far have been on the applied side (less so over time), and other modellers who prefer to do more basic research. There is also some correlation between positions on this question and the problem-oriented versus general-purpose controversy. Perhaps it is not humanly possible for applied and basic scientists in any field to learn to appreciate each other, but it would be nice if they did, because they are both necessary and depend on each other in many vital ways.

3. Should models be normative or descriptive?

Session 1 of the Conference (see Chapter 5) cleared up some of the semantic problems with this distinction. The discussion rose above the methodological differences between optimizers and simulators, and it was recognized that every model of every type has both normative and descriptive aspects. However, an important question about the most useful role of the modeller in society remained:

Modellers should create visions, design desirable futures, work out what is necessary to achieve worthy goals. They should not be negative prophets of doom, destroyers of the human spirit.

Modellers should analyse the imperfections of the current world and call attention to what will happen if these imperfections persist. They should not presume to set goals for society or to raise unrealistic expectations.

It is hard to imagine a model that does not, in fact, do both of these things, even if implicitly. A model of a desirable future in its very emphasis on what is desirable is pointing to what the modeller must believe are the faults of the current system. And a model that extrapolates the present into a messy future is nearly always put forth in order to change current behaviour so as to avoid the mess. Values abound in both approaches, but are easier to identify in the normative one. And descriptions of the modellers' view of the current system are also present in both, but are more easily examinable in the descriptive model. Why modellers seem to prefer one style of presentation over the other is probably a matter of genes, glands, and life experiences. It is very useful to have both types of modellers around.

4. How far into the future can one see with a model?

Not very far at all, and our confidence in what we see goes down rapidly as we look forward in time. Probably five, ten, at most fifteen to twenty, years is the upper

How far we can see depends on the basic time constants of the phenomena we are looking at. For physical capital, population, resource, technology, and

limit for any sensible modelling study.

environmental questions, fifty to a hundred years is reasonable – and, in fact, necessary, because decisions taken now have impacts that will last that long.

This broad span of estimates arises from paradigmatic differences about what one is trying to see. For precise point predictions of rapidly moving entities, one cannot be very certain very far into the future. For general time trends or policy responses of quantities that move slowly, something useful (but not a point prediction) can be said about what will happen over the very long run.

5. What is the best method to use for global modelling?

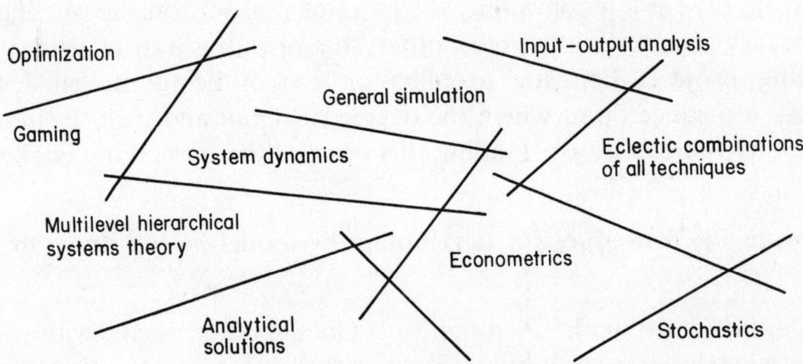

Most of the global modellers at the conference could have given firm, unhesitating answers to this question – but they would not have reduced to a simple bipolarity. This could be regarded as an amusing situation and everyone could be given a license to go off and play with his or her favourite toy, if it were not for the fact that methods shape both the inputs and the outputs of the models. Indeed, the method used is one of the best predictors of model structure and model results. Some process carried out by modellers, users, or both, is needed to assess the applicability of a chosen method to a given problem and to sort out the differing results arising from different methods. There are beginning attempts to do this – in the United States different energy models are being compared under similar scenarios – but untangling method from model from results is long, hard work.

6. Should models be large or small?

Big models get out of hand. They cannot be explained or documented. They cost a fortune to run. They require data that cannot be found.

Small models simply do not capture the real world. They give reasonable-looking, understandable advice that is totally wrong, because

They are black boxes full of undetectable errors. One should make small, transparent models and complicate with the greatest reluctance.

it is not based on all relevant factors. They are not credible to policy makers who know how complicated the world really is. The bigger the model the more comprehensive and reliable it is.

Onno Rademaker has summarized the discussion perfectly: 'Some people love simple models. Others love complicated models.' The difference usually traces back to the philosophical argument raised at the beginning of this chapter. Each group should probably go on making the best models it can, big and small, seeing as far as possible into the black boxes, and accounting as much as possible for omissions in earlier work.

However, even if one accepts a modeller's predilection for large or small models, there is still a continuing dilemma of realism, on the one hand, and transparency and feasibility, on the other, that produces a struggle in every new modelling project. For any problem or system being modelled there is probably a balance point where the degree of detail and inclusiveness is just enough and not too much. Finding this point is the continuing challenge.

7. Should the procedure for developing the model be top down or bottom up?

One should start with a rough working sketch, or even a running model, of the *entire structure* one is trying to represent. This way the big interactions and the interfaces among subsectors will be clear and nothing important will be omitted. Then one can elaborate, add detail, estimate parameters, and polish the model until time or money run out.

One should begin with complete models of all the *separate subsectors*. They should be carefully estimated, tested, elaborated, and explored, so that one understands and has confidence in each part of the model before they are fitted together. Each sectoral model can then be of use even before the pieces are combined.

Top-down people tend to be simple-model people and bottom-up people tend to be complicated-model people, but the correlation is not perfect. We know of several dramatic conversions to top-down philosophy engendered by total frustration in trying to assemble and understand the final model in a bottom-up project. We also know of some vehement advocates of top-down modelling who are at this time engaged in some of the most ambitious bottom-up efforts ever attempted. We can think of only a few suggestions that might help clear up this controversy:

- No modeller or modelling institution favouring either side should impose its preference on anyone else.

- Everyone who feels strongly in favour of one procedure should try the other at least once, just to see what happens.
- Someone should try a project that uses both strategies simultaneously.
- Groups who feel confident about using one strategy or the other should keep records and write honest reports about the progress of their work, the difficulties encountered, the triumphs, and the advice they would give about how to do better next time.

8. What should be done when data about a crucial system relation are not available?

Guess.	Leave it out.

There are excellent arguments to back up both of these positions. They permeate the rest of this book, and so there is no need to repeat them. Like most of the other issues discussed here, the side one favours may well be a matter of personal style overlaid with a veneer of scientific-sounding rationalizations. It would be fascinating to take two models that exemplify the extreme positions and investigate the information sources underlying each model's basic assumptions (methodological, structural, boundary). We suspect that each model would exhibit a great deal of both guessing and leaving out, although the areas in which the modellers chose to do one or the other would differ greatly.

However, within each basic position there is also a balance-point problem that faces each modeller anew in every modelling effort. Data are somehow never sufficient for what any modeller really would like to do. To what extent do you give up your aspirations and limit your model because of unavailable information? To what extent do you accept and manipulate defective, fuzzy, or anecdotal information to include aspects of a system that you think are important? To what extent do you make bold guesses? How do you know when you have gone just as far as the information can justify and no farther? No modeller of any philosophical school finds these questions easy. No universal formula can be followed to find the answers. They are different for every modelling project.

What we can do is be more understanding, humble, and tolerant of each others' struggles at the edge of the unknown. We can also provide a somewhat warmer professional climate for those who try the bolder experiments. Even experiments that fail teach all of us something.

9. How should actors, technology, prices, population, etc. be represented?

Actors

Aggregate them by function.	Don't aggregate them at all.

Represent their actual decisions exactly as they make them, including their goals and dissatisfactions.	Represent what you know to be the net result of their decisions (usually an optimization of something).
Only national governments are important actors.	There are important actors on every level, from individuals to multinational organizations.
Technology	
It's unpredictable. Assume it is constant.	It's unpredictable. Assume it makes any problem easier to solve exponentially over time.
It is a general, increasing tendency of an economy to use its resources more and more productively.	It is very specific to certain sectors and problems, it progresses in spurts, it costs money, takes time, and produces bad side effects.
It is the most important determinant of the future.	It is really pretty irrelevant to the future.
Prices	
Any model that doesn't include them is just plain stupid.	Prices are only information links regulating flows of real stuff. All you need to model is the real stuff, not the prices.
Prices are vital to the efficient distribution of scarce goods; they are the signals of scarcity.	Prices are politically determined to suit the interests of the ruling classes; they disguise real scarcity.
Population	
Birth rates are unpredictable; keep population exogenous, preferably using uncontroversial official UN extrapolations.	Making population exogenous implies that we must adapt to population growth rather than control it. Emphasize the endogenous influences on birth rate.
Avoid antinatalist formulations;	The main role of population in the

emphasize that if you take care of development, population will take care of itself.

global system is to drain economic output away from investment and into consumption.

Etc.

Surely there can be no one 'right' way to represent any of these factors. The choice depends on the system, the problem, the method, the data, and the ideology and values of the modeller. May a thousand flowers bloom.

10. How should a model be tested?

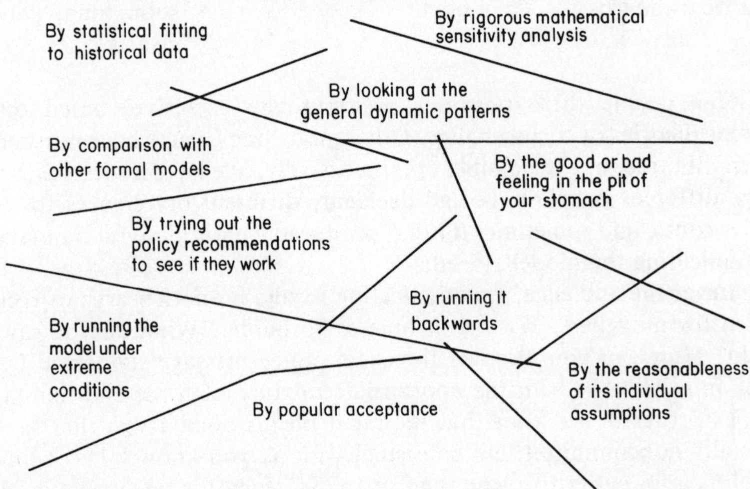

What kinds of tests should be used for what purposes is a question that can produce long and barren discussions. However, there is one point of agreement (reported later in this chapter): no one is satisfied with the testing that the global models have received so far. In some cases the dissatisfaction is because 'someone else's model has not been tested *my* way'. In other cases people are angry because someone took a model made within one methodological paradigm and applied tests from another that are considered inappropriate, misleading, and distorting by the first. There is hardly any subject within the field of social-system modelling that generates so little understanding and so much intolerance. Perhaps the most important statements to listen to are the modellers talking about the testing they have done on their own models by their own standards, why these tests were or were not satisfying, and what they see as obstacles to proper testing as they define it. More on this subject later.

11. What is the appropriate audience for global modelling? When and how should results be communicated to this audience?

The audience is the scientific community. The results should be presented in full technical detail for evaluation before any decision can be made about policy application. Results should be evaluated by scientific standards.

The audience is the current corporate and/or national leaders. They should be involved in the modelling process as directly as possible, preferably in private briefings and on-line interaction with the model.

The audience is the general public. The other two audiences suggested here should be avoided because they will contaminate the model with establishment biases and pressures. Results should be presented soon, loudly, and clearly.

For some people these positions are mutually exclusive, based on very different theories of social change. Other modelling groups have pursued two, or even all three, of the possible options actively. We have noticed that *within* groups different people have had decidedly different preferences for one or another route, and sometimes this has produced a natural division of labour in communicating the model's results.

Whatever the audience, communicating results is another area of recurring anguish for modellers. When is it time to go public? What do you say when you do? How can you discover the exact policy message supported by your model and deliver it with the appropriate mixture of force and humility? A statistics professor we know has mounted on his door a sign that says, 'Be maximally noncommittal, but consistent with all you know.' This counsel of caution is a lot easier to recite than to follow, even for one's own modelling project – and it is impossible to tell someone else how to do it.

This is the Way Modelling Should Be: Areas of Agreement

1. It is better to state your biases, insofar as you are able, than to pretend you do not have any.

There was total agreement at the Conference that no one making a global model (or any other kind of model, for that matter) can possibly be free from biases induced by cultural and ideological backgrounds, academic training, and methodological preferences. The most interesting and important biases are the ones that are not even quite conscious, the ones that only can be detected as vague uneasinesses that appear whenever economists talk to ecologists, or linear programmers to system dynamicists, or big-model people to small-model people, or capitalists to Marxists. One can never be fully aware of all one's biases. But one can be aware that one is unaware and deliberately expose oneself to situations in which biases can be discovered (the best way is to 'hang out' with whomever one regards as the enemy). In the presence of all the global modellers, the special preferences and phobias of each different global modeller become evident.

Given the thorough agreement that modelling is not and can never be a detached, totally objective activity, it would be a refreshing change to see modellers present themselves in person and in print as involved human beings instead of detached, objective mechanics.

2. Computer models of social systems should not be expected to produce precise predictions.

Most people think that social systems at some level are inherently unpredictable, either because human beings have free will or because the prediction itself is part of the system being predicted. Whatever the ultimate possibilities, at present we are far from being able to predict social-system behaviour, except perhaps for carefully selected systems in the very short term. Effort spent on attempts at precise prediction is almost surely wasted, and results that purport to be such predictions are certainly misleading.

On the other hand, much can be learned from models in the form of broad, qualitative, conditional understanding — and this kind of understanding is useful (and typically the only basis) for policy formulation. This conclusion is important enough to list separately.

3. Inexact, qualitative understanding can be derived from computer models and can be very useful.

If your doctor tells you that you will have a heart attack if you do not stop smoking, this advice is helpful, even if it does not tell you exactly *when* a heart attack will occur or how bad it will be.

In this vein, here are some examples of what we consider to be useful — albeit imprecise — statements emerging from global models so far:

- Reducing food consumption in rich countries will not reduce world hunger; in fact, it might increase the number of hungry people. A series of policies in the rich countries that will stabilize world food prices at relatively high levels is likely to reduce hunger.
- One of the most sensitive variables determining the growth rate of an economy is the lifetime of its industrial capital. Increasing the capital lifetime enhances economic growth.
- Continuing the present world trade and aid patterns will not close the income gap between the rich and poor countries.
- Natural delays in the dispersal of long-lived pollutants mean that pollution levels in the biosphere can go on rising for decades after pollution-control measures are put into effect.

- Even with the most optimal allocation of economic resources, economic growth and meeting basic human needs in several areas of Asia will be exceedingly difficult over the next few decades.
- Intermediate and gradual oil price rises are better for both consumer and producer nations than either very high or very low fixed prices.

4. Methods should be selected to fit problems (or systems); problems (or systems) should not be distorted to fit methods.

Each modelling technique was evolved originally to answer a particular type of question or study a particular kind of system. Each is admirably suited for some purposes and not others. They are like different tools in toolbox – some best for pounding in nails, some for sawing wood. It would be nice if each modeller were skilled in the use of all tools and as capable as a carpenter of picking up the right tool for each purpose. Unfortunately, these tools take years to learn to use well, and each comes together with a pair of glasses that influences how the user sees the world. As Maslow says, 'If your only tool is a hammer, you treat everything as if it were a nail.'

The match between purposes and methods in global modelling has not been very exact so far. Many of us have been pounding in nails with saws. Problem definition should precede method selection, and we should all have the integrity to turn down opportunities for modelling that do not match our favourite methods. Or, to put it another way, if you drop your key in a room where there is a light, switch on the light. Don't go outside to look just because you like streetlights.

5. The most important forces shaping the future are social and political, and these forces are the least well represented in the models so far.

There is no mention in any of the models described in this book of armaments or wars or multinational corporations, of new ideas, of political leaders or parliaments or terrorists, of religions or communications or advertising or spiritual awakenings, of families or love or generosity, or the influence of values on historical development, or vice versa. Every global modeller knows that at least some of these things exist and are important. Many have deliberately set out to include at least some of them in their models. Yet somewhere in the crunch of getting the model made, the political scientists and sociologists are not listened to, and the 'soft stuff' gets left out. At the Sixth Global Modeling Conference just about everyone, including those we would label as the most hard-nosed mathematicians, deplored these omissions.

Why do 'soft' variables get left out?

The commonest explanation is that there are not sufficient theories and/or data to include sociopolitical variables. The most poignant explanation that surfaced at the Conference was, 'We thought that, if we put that stuff in, all you guys would laugh at us!' Neither of these is really a good reason.

6. In long-term global models, environmental and resource considerations have been too much ignored.

Only two of the seven global models discussed at the Conference contained any representation of environmental factors, and even these two modelling teams did not feel satisfied with their environmental sectors. The problem here can hardly be lack of theories or data, since the field of ecology is at least as well developed as that of economics. But somehow there has been little communication between the ecologists and the global modellers so far.

7. Models should be tested much more thoroughly for agreement with the real world, for sensitivity to uncertainties, and over the full range of possible policies.

Although the various schools of modelling have sharply differing ideas about how to perform tests, it was agreed that global models generally have been poorly tested, even by the standards of their own schools (with some notable exceptions). The one test that every model passed, agreement with historical data, was not viewed as very impressive by anyone. Interesting suggestions for better testing included these:

- Give the model to an independent or opposing group for testing.
- Be sure that all policy tests are symmetrical (if you test for lower trade barriers, also test for higher ones; if you raise estimates of resource availability, also lower them; if you assume technology will improve faster, also assume it will improve more slowly).
- At the beginning of the project, appoint someone who cannot be fired to be the resident critic and model tester.
- Throughout the project keep a notebook in which all doubts, uncertainties, guesses, and alternative formulations are entered. Refer back to this notebook and test each uncertainty for its effect on the model results.
- Be sure the model sponsor appreciates the importance of testing and budgets sufficiently for it.
- Test for sensitivity with relation to the real system; it is of no interest just to know whether the model is sensitive; however, it is very important to know if it is sensitive where the real system is sensitive, and insensitive where the real system is stable.
- Test for whether or not the *policy conclusion* is sensitive; it may be uninteresting to know that a 5 per cent change in x produces a 10 per cent change in y, but it is vital to discover whether small changes in x can reverse your policy advice.

In the entire field of social-system modelling, there are no clear rules for producing an overall evaluation of a model. Global modellers are as confused in this area as everyone else.

8. A substantial fraction of modelling resources should go to documentation.

9. Part of the model documentation should be so technically complete that any other modelling group could run and explore the model and duplicate all published results.

10. Part of the documentation should be so clear and free from jargon that a non-technical audience could understand all the model's assumptions and how these assumptions led to the model's conclusions.

These three recommendations arose from the general agreement at the conference that the state of documentation in the field of social-system modelling in general is atrocious and that the global models have not set a shining example. There may be some good reasons why some kinds of models need not be documented (proprietary, fast turn-around times, etc.) or why scientific models do not need popular documentation. But for global models made for policy purposes there is no excuse for either secrecy or lack of clarity. No field that is supposed to be scientific can possibly progress if its practitioners cannot build on or criticize each others' work. No field that is supposed to influence decision making can be of use if the decision makers cannot understand it.

The few groups who have produced adequate documentation reported uniformly that documenting the model took at least as many man-years as making it. One reason models do not get documented is that this much effort does not get budgeted in advance for this purpose, so the money runs out before it is done. Another reason is that documentation is tedious, and the enthusiasm runs out before it is done. Both problems can be avoided if documentation is an ongoing, constantly updated activity, instead of one nobody thinks about until the fun is over.

11. Modellers should identify their data sources clearly and share their data as much as possible.

Problems with data are bad enough without having to reinvent the wheel. We should not only support each other by sharing what data we have unearthed, we should also perhaps clarify where we have rejected unacceptable data or searched to no avail, to save each other some digging.

12. Users, if there are any clearly identifiable ones, should be involved in the modelling process as directly and frequently as possible.

There are two reasons for this. First, the interested attention of a user is likely to keep the model true to its purpose and reflective of real-world conditions and constraints. Second, a user who has helped shape a model and who understands it is likely to implement its results. Neither of these rules is helpful to modellers who view the general public as their audience. There may be some good rules about public presentation, dealing with the media, etc., but no one at the Conference had any such rules to suggest.

13. It is necessary to have an international clearinghouse for presenting, storing, comparing, criticizing, and publishing global models.

IIASA has played this role admirably in the past, and should continue to do so in the future.

14. There should be many more global models.

Major global problems seem to be increasing and to be increasingly poorly understood. Global models have only begun to shed even the dimmest light on some of these problems; some of the most urgent have not yet even been addressed. As long as there is one undivided planet around which matter, energy, technologies, and information flow, there will be a need for analysis that transcends national interests and deals with problems beyond the space and time horizons of political institutions. The scope of an analysis must match the scope of the problem.

GLOBAL PROBLEMS REQUIRE GLOBAL MODELS.

A final word about making models

'Technology is a crucial factor in global development.'
'How do you know that?'
'Well, I don't know it, but all the global models assume it; I guess it's really an act of faith.'

Here are some other acts of faith that regularly appear in global models of both the mathematical and mental kinds:

1. The poor nations of the world are developing in the same pattern as the Western industrialized nations, but after a time lag.
2. Political leaders are above the global system, outside it, making the important decisions affecting it, and not affected by it.
3. The most important phenomena in the world are economic and can be described in terms of monetary units.
4. Nation-states are the basic actors in the world (not multinational corporations, not media, not households), and they interrelate primarily through flows of commodities and money.
5. A good indicator of the welfare of a population is the annual flow of market-exchanged goods and services (measured by monetary value) produced by that population.
6. The questions that political leaders are interested in are indeed the critical questions that need to be answered.
7. The questions that political leaders are interested in are narrow and self-serving and not the important ones at all.

It could be that some of these statements are true.
Or that some are useful approximations for some purposes.
Or that some or all are totally false.
Or, most disturbing of all,
that by stating and believing in them, we make them true.

Should any of them be accepted unquestioningly as the basis for our global models?

conceptual walls	Assumptions like these are That limit our thinking.	conceptual walls

They interfere with seeing the world as it really is
and with imagining the world as we really want it to be.

Why do they dominate our models?

Where do our models, mathematical and mental, come from?

How do they get walled in and unimaginative?

The environment for modellers
of all types
is not very open or safe.

We do not feel free to put forward creative new hypotheses that might affront conventional wisdom.

We are not eager to put our hypotheses to stringent tests against the real world.

We are only rewarded for being right.
We are laughed at for being wrong.
So we design our tests and select our results to maximize the possibility of being right.

We do not value bold ventures that teach us something because they didn't come out as expected.

We hide them
and miss the lessons in them
so we can look good.

We do not have a professional climate that permits us to report our failures honestly as well as our successes.

Human beings learn by trial and error
error
error.

It's very hard to learn if you never allow yourself to make an error.

It's very hard to teach if you never admit and draw lessons from your errors.

We could do a lot to increase the rate at which we learn from our models,
both in the way we do and report our own modelling and in the way we receive the modelling of others.

A final word about using models

Given all the difficulties, uncertainties, quirks of computer models discussed in this book,

What, if anything, should people, especially 'decision makers', do with/to/for/because of/in reaction to them?

Although we are computer modellers much of the time,
Sometimes we are decision makers too.

This is what we do:

1. We keep watching and listening, paying attention to the questions raised by the global models and the tentative answers put forth.
2. We ask a lot of questions of the modellers whenever we get a chance. We keep insisting that the answers be in plain, understandable English. When an answer comes that we don't understand, we say so and ask for a simpler phrasing. If we find a modeller who simply cannot communicate clearly, and if we can't find anyone who can translate that modeller understandably, we stop listening.
3. We try to probe for the biases of each modeller and correct for them (we would do that with any 'expert', computer modeller or not).
4. We look for and encourage several different models made by different people with different methods addressing the same problem. If possible we get them together to work out their differences while we listen and question.
5. We get information from lots of other sources in addition to the computer model, especially information we know is hard to incorporate in computer models (information about values and attitudes, about politics and factions, about the feelings and goals of those who will be affected by our decisions).
6. We remember that our decisions must come from incomplete, imperfect models in any case and that our mental models have a strong tendency to be biased and illogical.
7. We try to extract from the computer models general understanding, insights, appreciation for interrelatedness, which we can then apply in our own way to the problems before us. We would never expect the computer to tell us exactly what to do.
8. Ultimately, we rely on our own judgement, as we always have, hoping that that judgement has become a bit more informed and comprehensive by our interaction with the precision and consistency of the computer model.

Computer models do not make the job of decision making any easier.
They make it harder.

They enforce rigorous thinking
    and expose fallacies in mental models we have always been proud of.

We think the work is worth it.

We think it pushes our mental models to be a bit closer
    to reflecting
    the world as it is.

A final word about the globe

The most basic message of the global models is not new
and should not be surprising.

We do not need a computer model to tell us that:

we must not destroy the system upon which our sustenance depends.
poverty is wrong and preventable.
the exploitation of one person or nation by another degrades both the exploited and the exploiter.
it is better for individuals and nations to cooperate than to fight.
the love we have for all humankind and for future generations should be the same as our love for those close to us.

if we do not embrace these principles and live by them, our system cannot survive.

our future is in our hands and will be no better or worse than we make it.

These messages have been around for centuries.
They reemerge periodically in different forms
and now in the outputs of global models.

Anything that persists for so long and comes from such diverse sources as gurus and input-output matrices must be coming very close to

truth.

We all know the truth
at some deep level
within ourselves.
We have only to look honestly and deeply
to find it.

And yet we don't live as if we knew it.

Some of us actively deny messages like the ones from the global models.
Others try very hard not to think about them.
Most of us
feel helpless
shrug our shoulders
wish things were otherwise
assume that we can do nothing
and go on living.

Meanwhile, on this planet,
twenty-eight people starve to death each minute
one species of life disappears forever every day
and one million dollars are spent each minute on armaments.

The current condition of our globe is intolerable
and we make it so.

It is changing
because of what we decide.

It could be beautiful.
If we would only
decide to get along together
be open to each other and to new ways of thinking
remember what is really important to us
and what is less so
and live our lives for that which is important.

As sophisticated, skeptical, scientific Westerners
We always react to statements like that by saying

It sounds too simple
And is in fact impossible.
How could we ever decide to get along together?
You don't just decide things like that.
And how could we get everyone else to decide it?
(It couldn't be possible that everyone else is just like us
and is saying that same thing)

When everyone is so sophisticated
    that they can't believe it could be simple to be honest and to care

And everyone is so smart
    that they know *they* don't count
        so they never try

You get the kind of world we've got.

Maybe it's worth thinking another way

    as if we cared and we made a difference,

Even if it is just groping in the dark.

Appendix 1

The agenda of the conference

Tuesday, 17 October 1978

09:30-09:45	Welcome	R.E. Levien
09:45-10:00	Introduction to the Conference	G. Bruckmann
10:00-13:00	Topic 1: Purposes and Goals of Global Modeling	D.H. Meadows*
14:00-16:00	Topic 2: Methodological Problems	S. Cole,* P.C. Roberts*

Wednesday, 18 October

09:00-10:30	Topic 3: Policy Variables and Factors in Global Models	M.A. Keyzer*
11:00-12:30	Topic 4: Structural Aspects	O. Rademaker,* B. Moore,* M.A. Keyzer,* H. Linnemann*
14:30-16:00	Topic 5: Testing a Global Model	S.A. Bremer*
16:30-18:00	Topic 6: Internal Organization of Global Modelling Work	J.M. Richardson*

Thursday, 19 October

09:00-10:30	Topic 7: Relations between Modeler and User	B.G. Hickman,* J. Robinson*
11:00-12:30	General Evaluation	G. Bruckmann
14:30-18:00	Presentation and Discussion of New Models	

*Introductory statements

Friday, 20 October

09:00-12:30 and 14:30-18:00	Presentation and Discussion of New Models (continued)

Appendix 2

The list of participants

AUSTRIA

Dr. Peter Ahammer
University of Economics
Franz-Klein-Gasse 1
A-1190 Vienna

Dipl. Ing. Werner Horn
Department of Medical Cybernetics
University of Vienna
Schwarzspanierstr. 17
A-1090 Vienna

Dipl. Ing. Karl Heinz Kellermayer
Chair of Systems Theory
University of Linz
Altenbergstrasse
A-4045 Linz-Auhof

Dipl. Ing. Dr. F. Landler
Institute for Socio-Economic Development Research
Fleischmarkt 20
A-1010 Vienna

Dr. Kurt Mauler
University of Economics
Franz-Klein-Gasse 1
A-1290 Vienna

Prof. J. Millendorfer
STUDIA
Berggasse 16
A-1090 Vienna

Prof. Paul Neurath
Institute for Sociology
Alserstrasse 33
A-1090 Vienna

Dr. Franz Pichler
Department of Statistics and Computer Sciences
University of Linz
Altenbergstrasse
A-4045 Linz-Auhof

Dr. Ingo Schmoranz
Institute for Advanced Studies
Stumpergasse 56
A-1060 Vienna

Dr. Jiri V. Stolka
Austrian Institute for Economic Research
P.O. Box 91
A-1103 Vienna

Dr. Walter Buchstaller
Department of Medical Cybernetics
University of Vienna
Schwarzspanierstrasse 17
A-1090 Vienna

BELGIUM

Prof. Jacques Federwisch
Societe de traction et d'electricite
Rue de la Science 31
1040 Brussels

BULGARIA

Mr. Rumen Dobrinsky
Institute of Social Management
Ul. Pionerski Pat 21
Sofia

BRAZIL

Dr. Guillermo A. Albizuri
S.B.I. — Group ASDELA
Candido Mendes University
Rue Visconde de Piraja 351
Andar 7^0
Rio de Janeiro

Dr. Neantro Saavedra-Rivano
S.B.I. — Group ASDELA
Candido Mendes University
Rue Visconde de Piraja 351
Andar 7^0
Rio de Janeiro

Dr. Carlos A. Ruiz
S.B.I. — Group ASDELA
Candido Mendes University
Rue Visconde de Piraja 351
Andar 7^0
Rio de Janeiro

Dr. Hugo D. Scolnik
S.B.I. — Group ASDELA
Candido Mendes University
Rua Visconde de Piraja 351
Andar 7^0
Rio de Janeiro

CANADA

Prof. Paul Medow
York University
4700 Keele St.
Toronto

CSSR

Dr. Jiri Sliva
Institute of Philosophy and Sociology
Academy of Sciences of the CSSR
Jilska 1
Prague

FINLAND

Dr. Pol. Sc. Pekka Kuusi
The Finnish Committee for IIASA
Oy Alko Ab
Pl 350
SF-00101 Helsinki 10

BERLIN (WEST)

Dr. habil. Helmut Arnaszus
Institute for Philosophy
Free University of Berlin
Gelfertstrasse 11
D-1000 Berlin (West) 33

Dr. Stuart Bremer
Science Center Berlin
Steinplatz 2
S-1000 Berlin (West) 12

Dr. Thomas Cusack
International Institute for
Comparative Social Research
Science Center Berlin
Steinplatz 2
D-1000 Berlin (West) 12

Dipl. Math. Peter Otto
Science Center Berlin
Steinplatz 2
D-1000 Berlin (West) 12

Dr. Brian Pollins
Science Center Berlin
Steinplatz 2
D-1000 Berlin (West) 12

Dr. Christian Riethmüller
IFZ — Institute for Future Research
Technical University of Berlin
Giesebrechtstrasse 15
D-1000 Berlin (West) 42

Dr. Ulrich Widmaier
Science Center Berlin
Steinplatz 2
D-1000 Berlin (West) 12

Mr. Peter Haas
Free University of Berlin
Gelfertstrasse 11
D-1000 Berlin (West) 33

FEDERAL REPUBLIC OF GERMANY

Dr. Hartmut Bossel
ISP — Institute for Applied Systems Research and Prognosis School of Medicine
Haus W3G — Karl-Wiechert-Allee 9
D-3000 Hannover

Dr. Siegfried Dickhoven
Society for Mathematics and Computer Processing (GMD)
Schloss Birlinghoven
D-5205 St. Augustin 1

Dipl. Math. Wolfgang Oest
ISP — Institute for Applied Systems Research and Prognosis
School of Medicine
Haus W3G — Karl-Wiechert-Allee 9
D-3000 Hannover

GERMAN DEMOCRATIC REPUBLIC

Prof. Hans Knop
Berlin School of Economics
Hermann Dunckerstrasse 8
DDR-1157 Berlin

Prof. Manfred Peschel
Department of Mathematics and Cybernetics
Academy of Sciences
Rudawer Chaussee 5
DDR-1199 Berlin

Prof. A. Sydow
Central Institute for Cybernetics and Information Processes
Academy of Sciences
Rudawer Chaussee 5
DDR-1199 Berlin

HUNGARY

Dr. Zsuzsa Bekker
Natonal Planning Office
Arany Janos u. 6-8
Budapest V

Dr. Ernoe Zalai
Department of National Economic Planning
Karl-Marx University of Economic Sciences
Dimitrov ter 8
1093 Budapest IX

INDIA

Prof. Anirudh L. Nagar
Delhi School of Economics
University of Delhi
Delhi — 110007

ITALY

Dr. Giorgio Roncolini
Systems Analysis — Research Unit
Fiat Research Centre
Strada del Drosso 145
Turin

JAPAN

Prof. Yoichi Kaya
Electrical Engineering Dept.
Faculty of Engineering
University of Tokyo
Hongo 7-3-1, Bunkyolku
Tokyo 113

Prof. Akira Onishi
Department of Economics
Soka University
1-236 Tangi-machi
Hachioji-shi
Tokyo 192

Prof. Yutaka Suzuki
Department of Electrical Engineering
Osaka University
Yamada-Kami, Suita
Osaka

NETHERLANDS

Prof. Jerrie de Hoogh
Agricultural University
Wageningen
De Leeuwenborch
Wageningen

Dr. Michiel Keyzer
Center for World Food Studies
Free University Amsterdam
Amsterdam

Dr. Hans Linnemann
Free University Amsterdam
Amsterdam

Prof. ir. Onno Rademaker
Eindhoven University of Technology
Systems and Control Engineering Group
P.O. Box 513
Eindhoven 5600

Dr. W.J.M. Wils
Prinses Mariannelaan 246
2275 BN Voorburg

POLAND

Dr. Andrzej Dabkowski
Planning Institute
Zurawia 4a
00-507 Warsaw

SWEDEN

Dr. Anders Karlqvist
The Swedish Committee for IIASA
FRN
S-10310 Stockholm

SWITZERLAND

Dr. Gernot Ruths
Center for Economic Research
Swiss Federal Institute of Tehcnology
Weinbergstrasse 35
CH-8006 Zürich

Dipl. Ing. Jean-Michel Toinet
Institute of Energy Production
Swiss Federal Institute of Technology
17, Av. Dapples
CH-1007 Lausanne

UNITED KINGDOM

Dr. Sam Cole
SPRU
University of Sussex
Brighton

Mr. P.C. Roberts
Systems Analysis Research Unit
Department of the Environment
2, Marsham Street
London

USA

Prof. Anne P. Carter
Department of Economics
Brandeis University
Waltham, Mass. 02154

Prof. Bert G. Hickman
Department of Economics
Stanford University
Stanford, Ca. 94305

Dr. Frederick Kile
Aid Association for Lutherans
Appleton, Wisc. 54919

Mr. William T. Marin
al Dir'iyyah Institute of Geneva, Switzerland
1925, North Lynn St.
Arlington, Va. 22209

Prof. Donella Meadows
Dartmouth College
Hanover, New Hampshire 03755

Dr. Berrien Moore III
Center for the Study of Complex Systems
University of New Hampshire
Durham, N.H. 03824

Prof. John M. Richardson, Jr.
College of Public Affairs
Center for Technology and Administration
The American University
Washington, D.C. 20016

Dr. Michael D. Ward
Department of Political Science
Northwestern University
Scott Hall
1890, Sheridan Road
Evanston, Ill. 60201

USSR

Dr. Sergey Dubovsky
All-Union Institute for Systems Studies
Ryleyev Str. 29
Moscow 119034

Prof. Alexandr Granberg
Novosibirsk Institute of Economics and Organization of Industrial Production
Prospekt Nauki
Novosibirsk

Prof. Vladilen Clockoff
All-Union Institute for Systems Studies
Ryleyev Str. 29
Moscow 119034

Prof. Nicolaj Lapin
All-Union Institute for Systems Studies
Ryleyev Str. 29
Moscow 119034

INTERNATIONAL ORGANIZATIONS

EURATOM — ECE
Ispra 21020
Italy

Dr. Malcolm Slesser
Systems Analysis Divison
Joint Research Centre

ILO — International Labour Organisation
CH-1211 Geneva 22
Switzerland

Mr. Michael Hopkins
Population and Labour Policies Department

UNITED NATIONS
New York, N.Y. 10017
USA

Mr. Douglas Walker
Centre for Development Planning, Projections and Policies
Department of International Economic and Social Affairs
IESA/CDPP

UNITAR — United Nations Institute for Training and Research
New York, N.Y. 10017
USA

Prof. Vladislav Tikhomirov

UNIDO — United National Industrial Development Organization
Lerchenfelderstrasse
1080 Vienna, Austria

Mr. Eoin Gahan
Mr. Young Cho
Mr. Yuri Zelentsov

WORLD BANK
Washington, D.C.,
USA

Dr. Syama Prasad Gupta
Economic Analysis and Projects Department

IIASA

C. Almon
System & Decision Sciences

A. Andersson
General Research

T. Balabanov
Energy Systems

G. Bruckmann
General Research — Global Modeling Review

A. Bykov
Secretary to IIASA

V. Chant
Energy Systems

C. Csaki
Food & Agriculture

C. Clemens
Rapporteur

J. Eddington
Management & Technology

G. Fischer
System and Decision Sciences

P. Fleissner
Human Settlements & Services

K. Frohberg
Food & Agriculture

P. de Janosi
System & Decision Sciences

A.M. Khan
Energy Systems

R. Kulikowski
General Research

R. Levien
Director

H. Maier
Management & Technology

C. Marchetti
Energy Systems

D. Nyhus
Systems & Decision Sciences

R. Pestel
Management & Technology

A. Propoi
Systems & Decision Sciences

F. Rabar
Food & Agriculture

J. Robinson
Food & Agriculture

A. Timman
Food & Agriculture

R. Tomlinson
Management & Technology

R. Venturelli
Resources & Environment

A. Wierzbicki
System & Decision Sciences

I. Zimin
Energy Systems

Acknowledgments

Many people aided the creation of this book by providing encouragement, financial support, comments, criticism, and logistical help. Not all of them, perhaps, would like to be associated with the product, but we would like to thank them all, even those whose advice we did not follow.

There are, of course, the participants in the conference, whose contributions made the book possible — and many others too numerous to mention. But we must mention with special appreciation Ilse Brestan-Ziegler, Barbara Hauser, Yoichi Kaya, Roger Levien, Nino Majone, Hugh Miser, Ferenc Rabar, Bonnie Riley, Peter Roberts, Jenny Robinson, Andrei Rogers, Edward Quade, Hugo Scolnik, and Harry Tollerton.

Index